Dying Planet
The Extinction of Species

Jon Erickson

FIRST EDITION
FIRST PRINTING

Library of Congress Cataloging-in-Publication Data

Erickson, Jon, 1948–
Dying planet : the extinction of species / by Jon Erickson
p. cm.
Includes bibliographical references and index.
ISBN 0-8306-6726-1 (hardbound) ISBN 0-8306-7726-7 (pbk.)
1. Extinction (Biology) 2. Evolution (Biology) I. Title.
QE721.2.E97E965 1991
575'.7—dc20 91-9277
CIP

TAB Books offers software for sale. For information and a catalog, please contact TAB Software Department, Blue Ridge Summit, PA 17294-0850.

Acquisitions Editor: Roland S. Phelps
Book Editor:Andrew Yoder
Production: Katherine G. Brown
Series Design: Jaclyn J. Boone
Cover Design: Lori E. Schlosser
Cover Illustration: Denny Bond, East Petersburg, PA.

Contents

Other TAB books by the author:

Volcanoes and Earthquakes
Violent Storms
The Mysterious Oceans
The Living Earth
Exploring Earth From Space
Ice Ages
Greenhouse Earth
Target Earth

Acknowledgments

The author wishes to thank the following organizations, which supplied photographs for this book: the Department of Agriculture Soil Conservation Service, the National Aeronautics and Space Administration, the National Forest Service, the National Museums of Canada, the National Oceanic and Atmospheric Administration, the National Park Service, the U.S. Coast Guard, the U.S. Navy, and the U.S. Geological Survey.

Introduction

We are rapidly filling the world with ourselves. As a result, the rest of the living world is forced aside to make room for more of us. The Earth is rapidly losing valuable plant and animal life. If trends continue at this rate, by the middle of the next century the number of extinct species could exceed those lost in the great extinctions of the geologic past.

Historic extinction events were caused by natural phenomena. But today's extinctions, in which one hundred or more species are vanishing each day, are mainly caused by destructive human activities. As human populations continue to grow out of control and as mechanized destruction of large ecosystems continues at its present furious pace, tragic numbers of species will be forced into extinction.

Perhaps as many as four billion species have inhabited the Earth throughout geologic time. Estimates of the number of species living today range from about 5 million to perhaps 50 million. Most species escape detection and go about their lives completely unnoticed by man. Yet, many of these organisms play fundamental roles in food chains. If we wantonly destroy these organisms, we ultimately destroy ourselves because all life is linked in an interconnected biosphere.

Mass extinctions are important to the evolution of life, and when a major extinction event takes place, new species evolve to fill the spaces vacated by those that departed. During the last 570 million years, five major mass extinctions and five or more minor ones have occurred. All extinction events seem to have resulted from biological systems in extreme stress, brought on by a radical change in the environment.

There is also speculation that extinctions might be periodic, brought on by celestial phenomena, such as cosmic rays from supernovas or asteroid and comet impacts. The bombardment by a large asteroid is generally cited as the reason for the extinction of the dinosaurs and for 70 percent of all known species 65 million years ago. A similar asteroid impact 210 million years ago, which was responsible for the extinction of nearly half of the reptile families, might have resulted in the rise of the dinosaur species. Therefore, the dinosaurs might have been created and destroyed by asteroids.

Extinctions might also be caused by terrestrial phenomena, such as massive volcanic eruptions, whose ash clouds shade the planet, causing a rapid drop in global temperatures. Reversals of the Earth's magnetic field also coincide with extinctions. Some extinction events have coincided with the ice ages; the effects of lowered global temperatures on life are considerable. The living space of warmth-loving species is dramatically reduced to the tropics, reducing the habitat area and food supply and forcing large numbers of species extinct.

Major extinctions generally follow periods of climatic cooling. These lowered temperatures slow down the rate of biological activity, which in turn could affect species diversity. Then, new species would evolve that would be better adapted to the cold. Species that survive mass extinctions and have been around the longest, usually evolved during colder conditions. It is even speculated that our own ancestors, which evolved during the Pleistocene ice ages, were one of these new species.

The complex interrelationships among species and between species and their environments is still not fully understood. However, it is becoming more apparent that the destruction of large numbers of the world's species will lower their diversity and allow "pests" to flourish because their natural enemies have been destroyed. Therefore, the destruction of large numbers of species will leave this an entirely different biological world than the one mankind originally inhabited.

1

The Origin of Life

SOMETIME in the very obscured past, perhaps 18 billion years ago, the universe began with an initial singularity of infinite density and infinitesimally small volume. This creation seems almost too fantastic, so scientists have modified the "big-bang" theory somewhat to allow for a nonsingular beginning followed by a period of very rapid expansion, called *inflation.* Afterwards, there was a transition to the more conventional type of development that we observe in the universe today.

Within a few billion years after the big bang, the universe cooled to allow the formation of the simplest atoms, hydrogen and helium, which make up over 99 percent of all matter in the universe. These atoms clumped together to form billions of galaxies, each containing billions of stars. The stars in turn created all other elements, which would later become the building blocks of new stars and planets, including our own Sun and Earth (FIG. 1-1). In effect, we owe our very existence to the stars, which created an abundance of carbon atoms necessary for life (TABLE 1-1).

STELLAR CONTRIBUTIONS

Every atom in our bodies and all the minerals in the Earth were created by giant stars over a hundred times larger than the Sun. After a very hot existence, that lasted for only a few hundred million years, the giant star exploded and became a supernova. By the time the star reached the supernova stage, the

(COURTESY OF NOAA)

Fig. 1-1. Earth as seen from the GOES weather satellite.

nuclear reactions in the core became an explosive event. The star then shed its outer covering, while the core compressed into an extremely dense, hot body, called a *neutron star.* The supernova compression is comparable to condensing the Earth to about the size of a golf ball.

When a supernova spews vast amounts of stellar material into empty space, the material forms a nebula (FIG. 1-2) composed mostly of hydrogen and helium and particulate matter that comprises all the other known elements. Certain portions of the nebula are compressed by density waves passing through the galaxy. The compression causes the nebular matter to collapse into a *protostar.* Continued compression as a result of self-gravitation generates enough heat to initiate a thermonuclear reaction in the core, and the star ignites.

TABLE 1-1. The Abundance of Chemical Elements in the Universe

NAME	SYMBOL	ATOMIC NUMBER	ATOMIC WEIGHT	RELATIVE ABUNDANCE
Hydrogen	H	1	1	1.0 E 12
Helium	He	2	4	8.5 E 10
Carbon	C	6	12	4.8 E 8
Nitrogen	N	7	14	8.5 E 7
Oxygen	O	8	16	8.0 E 8
Fluorine	F	9	19	3.4 E 4
Neon	Ne	10	20	1.0 E 8
Sodium	Na	11	23	2.1 E 6
Magnesium	Mg	12	24	3.9 E 7
Aluminum	A1	13	27	3.1 E 6
Silicon	Si	14	28	3.7 E 7
Phosphorous	P	15	31	3.5 E 5
Sulfur	S	16	32	1.7 E 7
Chlorine	Cl	17	36	1.7 E 5
Argon	Ar	18	40	3.3 E 6
Potassium	K	19	39	1.3 E 5
Calcium	Ca	20	40	2.3 E 6
Scandium	Sc	21	45	1.3 E 3
Titanium	Ti	22	48	1.0 E 5
Vanadium	V	23	51	1.0 E 4
Chromium	Cr	24	52	4.8 E 5
Manganese	Mn	25	55	2.9 E 5
Iron	Fe	26	56	3.3 E 7
Cobalt	Co	27	59	7.8 E 4
Nickel	Ni	28	59	7.8 E 4
Copper	Cu	29	64	1.9 E 4
Zinc	Zn	30	65	5.0 E 4

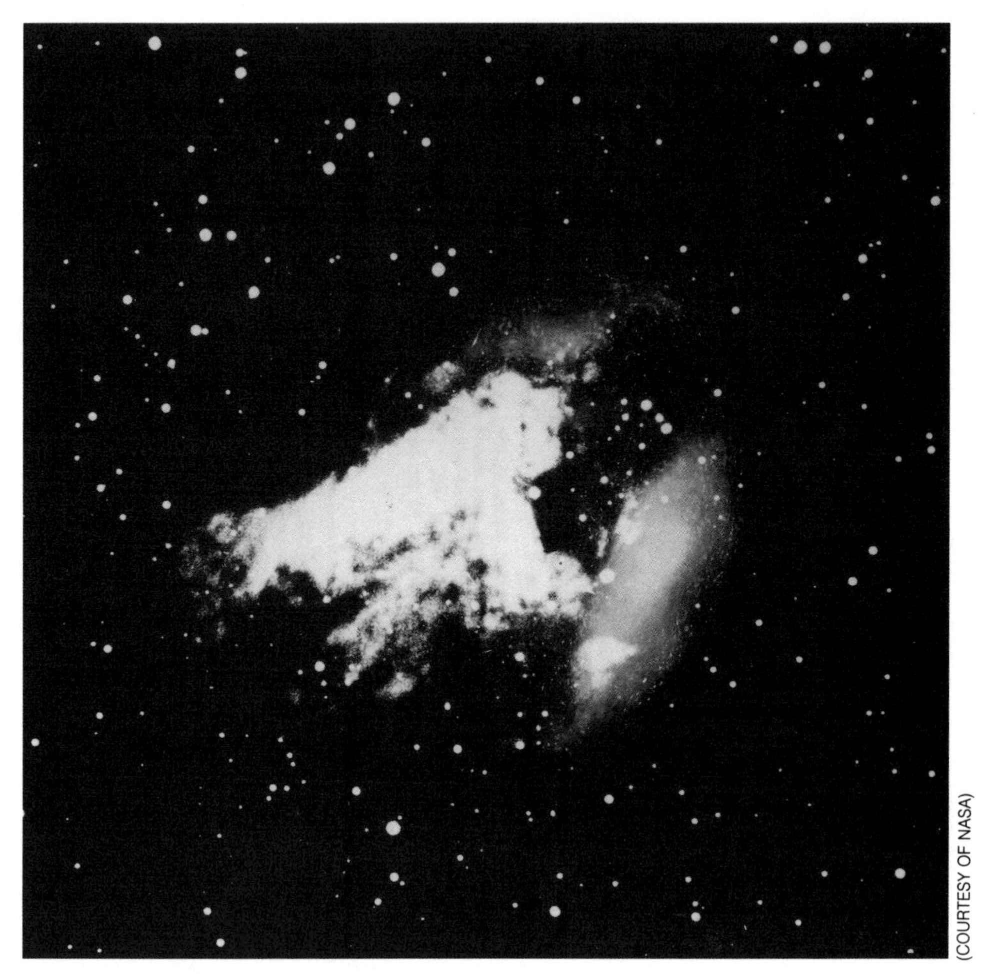
(COURTESY OF NASA)

Fig. 1-2. The Swan Nebula in Sagittarius.

A new star is born in our galaxy every few years or so. Stars originate from an assortment of nebulas, molecular complexes, and globules, composed of condensing clouds of gas and dust. A typical globule has a radius of about a light year and a mass of about 100 solar masses. Besides hydrogen and helium, galactic clouds contain an assortment of organic compounds, including carbon monoxide, formaldehyde, and ammonia.

About one million years is required for a solar nebula to collapse into a star. A density wave that originates from a nearby supernova provides the outside pressure needed to trigger the collapse. As the solar nebula collapses, it begins to rotate faster and faster. The spiral arms peel off from the rapidly spinning nebula to form a *protoplanetary disk.* The image of the original disk is still visible throughout our Solar System (FIG. 1-3) in the motions of the planets around the Sun.

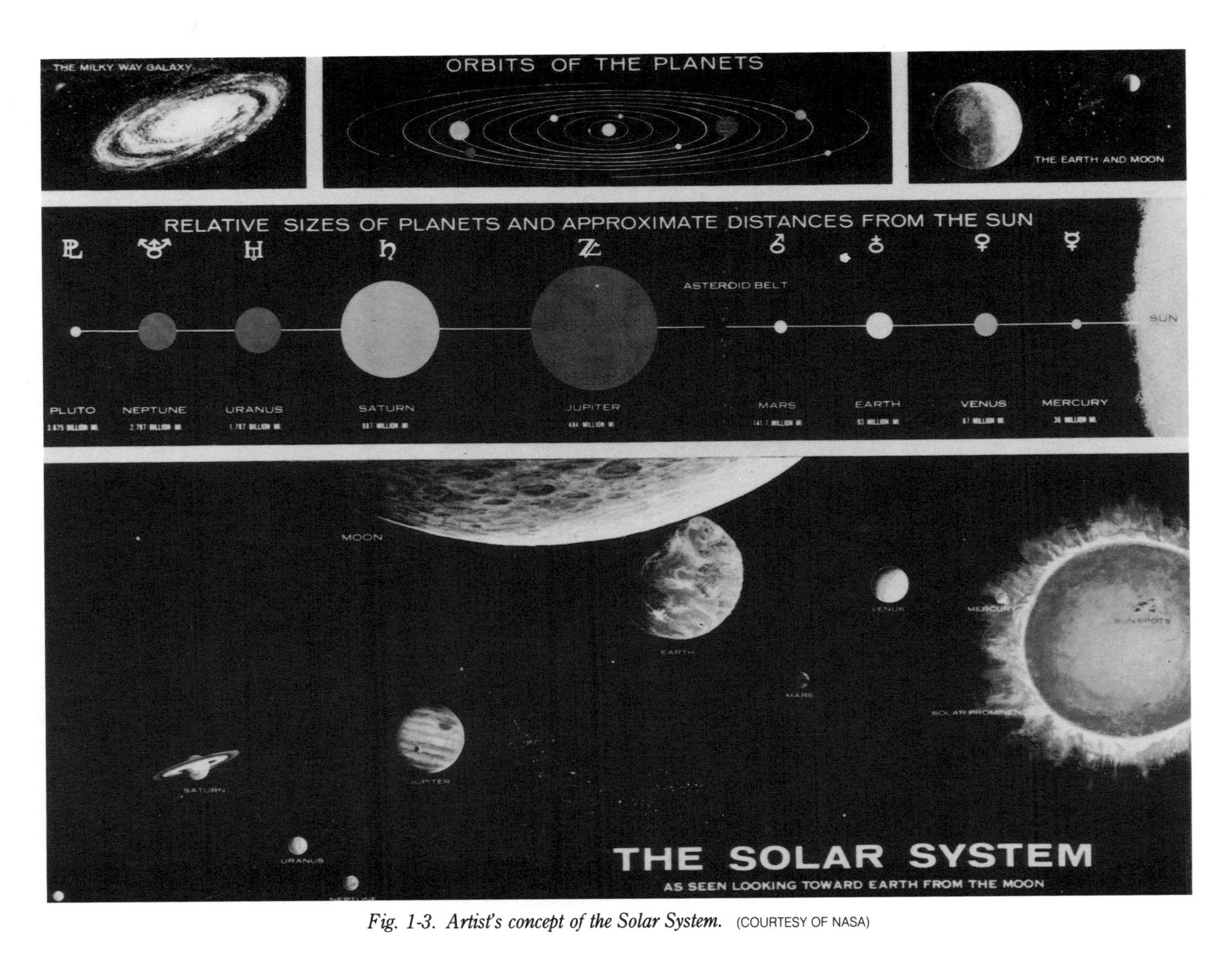

Fig. 1-3. Artist's concept of the Solar System. (COURTESY OF NASA)

Orbiting the Sun is a collection of planets and their moons; *asteroids*, which are pieces of broken planetoids; and *comets*, which are "dirty snowballs" formed from leftover gases and ices that were swept out of the main Solar System by the early Sun's strong solar wind. When the Sun first ignited, it produced intense particle radiation. The radiation pressure resulted in a solar gale, compared to the gentle breeze it is today. Much of the leftover debris revolves around the Sun in erratic orbits that can take them across the Earth's path. Every so often, a collision occurs that disrupts all life and causes mass extinctions.

PREBIOTIC LIFE IN SPACE

Venus might have once had oceans that covered the planet just as those on Earth. Spawned from the same material as the Earth and Mars, Venus should also have had a wet history. Yet today, the planet is essentially bone dry—with hardly a trace of water. It is postulated that life might have originated on Venus when the young Sun was dimmer and the planet was cooler and still possessed oceans.

For the first billion years or so of its life, the Sun produced about 30 percent less energy than it does now, and the amount of solar energy striking the Earth was the same as it is at Mars today (FIG. 1-4). Even with a lowered solar output, Venus' heavy carbon dioxide atmosphere produced a strong greenhouse effect that burned off all its surface water. With it evaporated any semblance of life. The

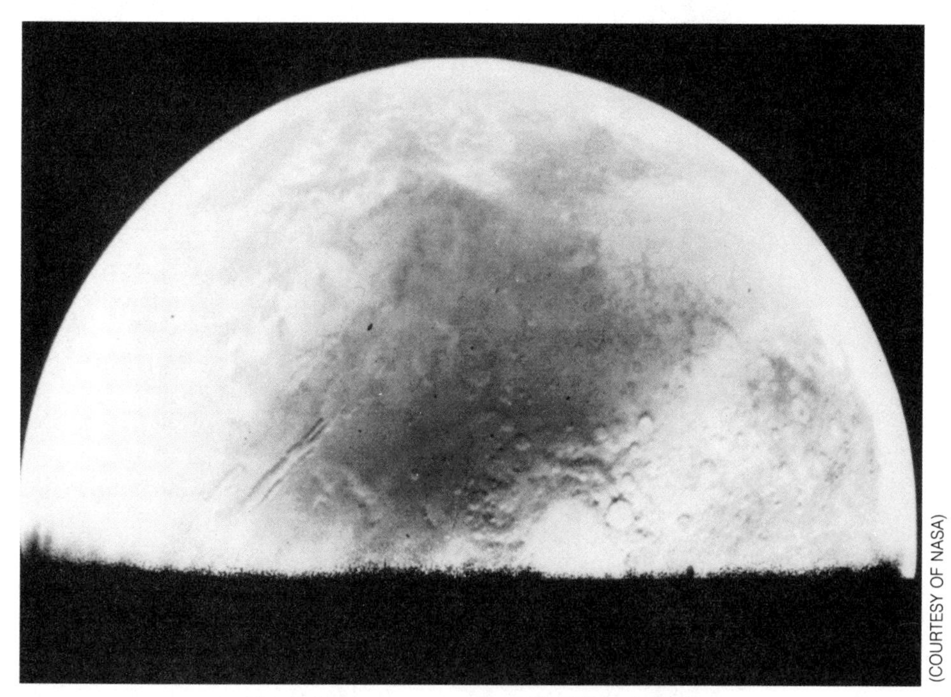
(COURTESY OF NASA)

Fig. 1-4. Mars as seen from the Viking 1 Orbiter.

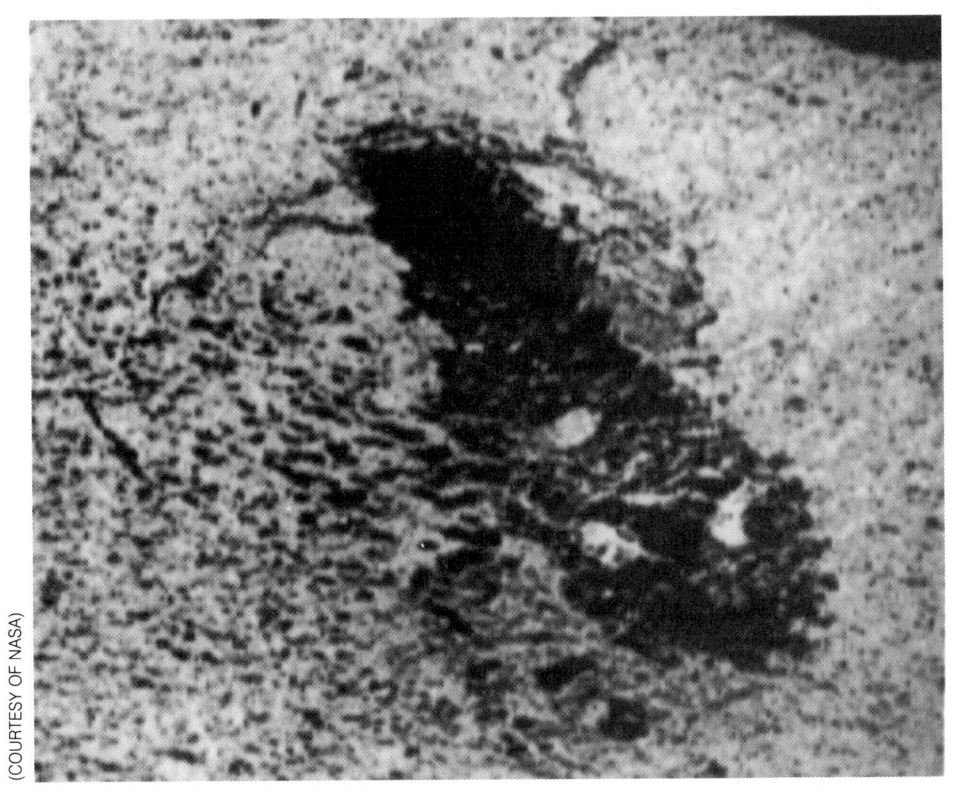
(COURTESY OF NASA)

Fig. 1-5. A trench excavated in the Martian soil by the Viking 1 surface sampler.

Earth might find itself in the same situation a billion years from now, since the Sun continues to increase in luminosity at a linear rate of about 1 percent every 100 million years.

The discovery of lichens in tiny pores on the undersides of rocks in Antarctica has fueled speculation that similar life forms might be found on Mars. Soil sampling conducted on the planet by the *Viking* landers in the late 1970s has indicated that the planet might have once harbored living entities. No direct evidence of life was found, however, mainly because the landers only sampled the surface sediments (FIG. 1-5), whose organic compounds might have been destroyed by strong ultraviolet radiation from the Sun. Nevertheless, signs of life might exist deeper into the crust. Instead of burning like Venus did, Mars became a frozen planet. Therefore, Mars would have to thaw before life could start again.

Scientists have long speculated that life might be possible in Jupiter's multicolored bands of clouds (FIG. 1-6), which are composed mostly of condensed water vapor and ammonia. The clouds are colored by organic substances brought up by strong convection currents. Jupiter's core, which is mostly made of rock and is about the size of the Earth, produces a large amount of internally generated heat. The heat that rises to the surface exceeds the amount of heat that the planet receives from the Sun. The organic substances in Jupiter's atmosphere interact with the Sun's ultraviolet radiation to produce a variety of organic compounds.

Space probes have searched the outer moons that are terrestrial in nature for any precursors of life. In November 1980, *Voyager I* was steered toward Saturn's largest satellite, Titan, in order to get a closer look at this most unusual of moons. Unlike all the other bodies in the outer Solar System, Titan has a thick primordial atmosphere that is even denser than the Earth's. The mix of gases in Titan's atmosphere is believed to be similar to that of the Earth's early atmosphere. Therefore, Titan is thought to be the best place to look for precursory signs of life.

Titan's atmosphere (FIG. 1-7) has retained conditions that are believed to have existed on all the planets soon after their formation. It contains carbon, nitrogen, and hydrogen, which are the essential ingredients for making amino acids. Methane is also abundant. However, the atmosphere contains no oxygen to interfere with the building of organic molecules. With large amounts of oxygen present, organic compounds would oxidize to form water and oxides of carbon and nitrogen. Under these reducing conditions in which oxygen is absent, the chemical reactions that occur in Titan's atmosphere might well give rise to some of the same organic molecules that are thought to have been the precursors to life on Earth.

(COURTESY OF NASA)

Fig. 1-6. Cloud bands in Jupiter's atmosphere from Voyager 1.

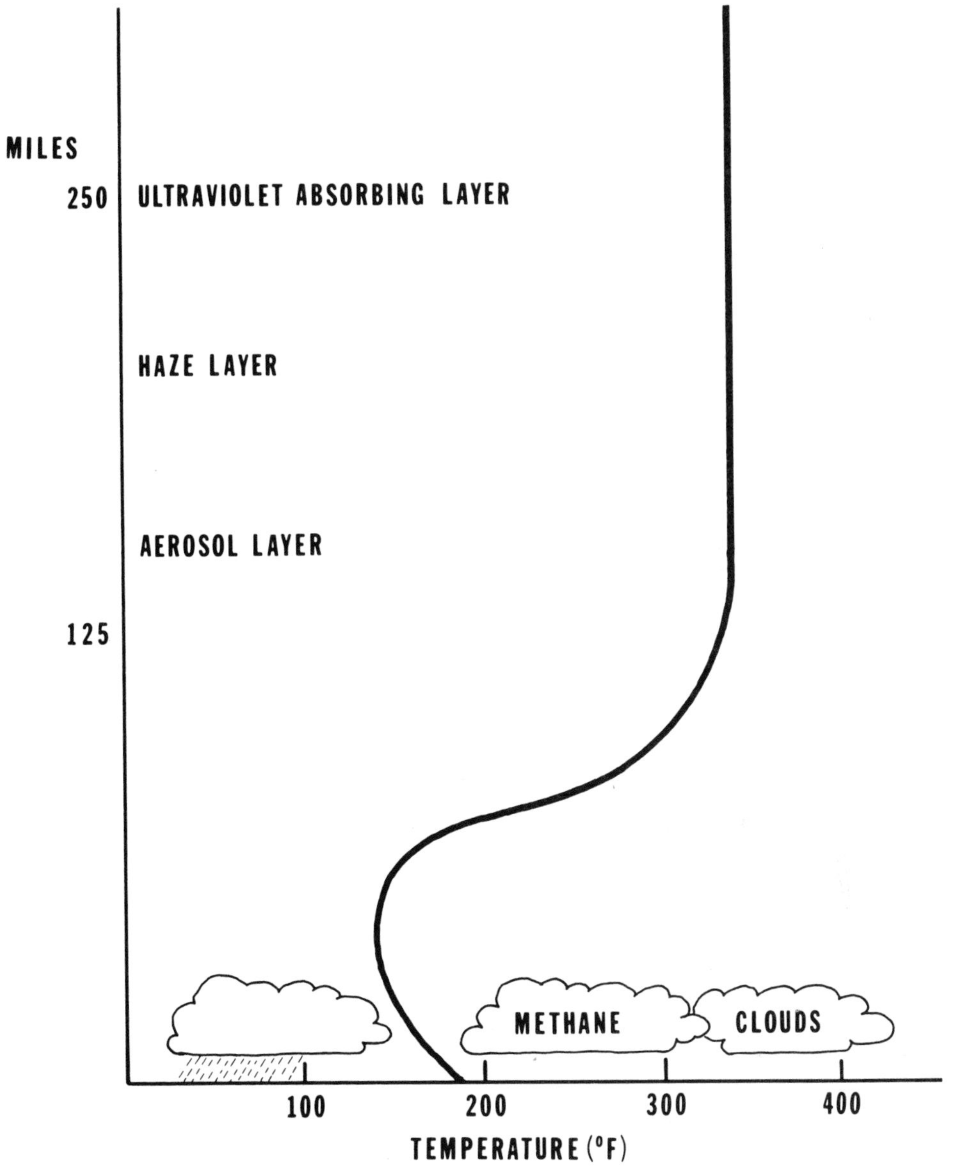

Fig. 1-7. The atmosphere on Titan is similar to Earth's early atmosphere.

Many speculate that life on Earth originally came from outer space and that the planet was seeded by meteorites or comets from other parts of the galaxy. The vehicle that organic substances could use to traipse around the galaxy are meteoroids composed with either a variety of space junk cast off by a supernova or an agglomeration of cosmic dust into small *planetesimals* or *planetoids.* Intragalactic dust clouds also contain organic molecules. The synthesis of organic molecules in the dark clouds might have been triggered by cosmic rays from deep space. These organic compounds could have been incorporated in the comets and meteoroids that pounded Earth in its early history.

The case for the existence of life outside Earth is compelling. If for no other reason, there has been plenty of time for life to evolve elsewhere in the galaxy, which is three times older than Earth itself. Nor was it necessary for life to evolve at a common point in space. It could have had several centers of creation. Just as it is thought that all parts of the universe seem to obey the same physical laws, so can it be said about biological laws. Therefore, the most fundamental definition of life everywhere in the universe would be that it is carbon-based, stores energy, contains genetic information, repairs itself, and makes near-exact copies of itself—passing its genetic blueprint on to its offspring.

The 4.5-billion-year-old Murchison meteorite discovered in Western Australia in 1969 is thought to be made up of much the same material that formed the Earth. Moreover, the interior of the meteorite contained lipidlike organic compounds that were able to self-assemble into cellularlike membranes: an essential requirement for the first living cells. The organic chemicals provided strong evidence for the existence of extraterrestrial amino acids. The material in the meteorite contains many of the essential components needed for creating life. This being the case, the Earth would not necessarily have been the only body in our Solar System, or in the entire galaxy for that matter, to receive the seeds of life.

THE PRIMORDIAL SOUP

Life is based on the chemistry of light elements, namely hydrogen, carbon, oxygen, and nitrogen. All amino acids, the very building blocks of life that make up molecules of proteins and DNA, have been synthesized by spark discharge experiments. These experiments used an apparatus to represent the early conditions on Earth (FIG. 1-8) with a variety of ingredients that are believed to have comprised the primordial atmosphere and ocean.

A flask of boiling water, representing the ancient seas, is connected to a large spherical vessel (above) that contains ammonia, nitrogen, methane, and water vapor, representing the early atmosphere. Inside the vessel, two electrodes provide a spark gap, through which high-voltage electricity is passed to stimulate powerful lightning bolts and strong ultraviolet radiation from the Sun. After passing through the spark gap, the gases condense and collect in a trap (below) that is connected to the boiling flask. Thus, gases and water vapor circulate in a closed loop system, representing the Earth's hydrologic cycle.

After the apparatus has operated for about one week, the dark soup is collected and chemically analyzed. The condensate is composed of an assortment of carbon compounds, among which are simple amino acids. This experiment proves that life was not a quirk of nature, but it instead followed certain logical rules by which organic chemistry operates. Therefore, given the conditions that existed on the early Earth, life would appear to have been inevitable from the start.

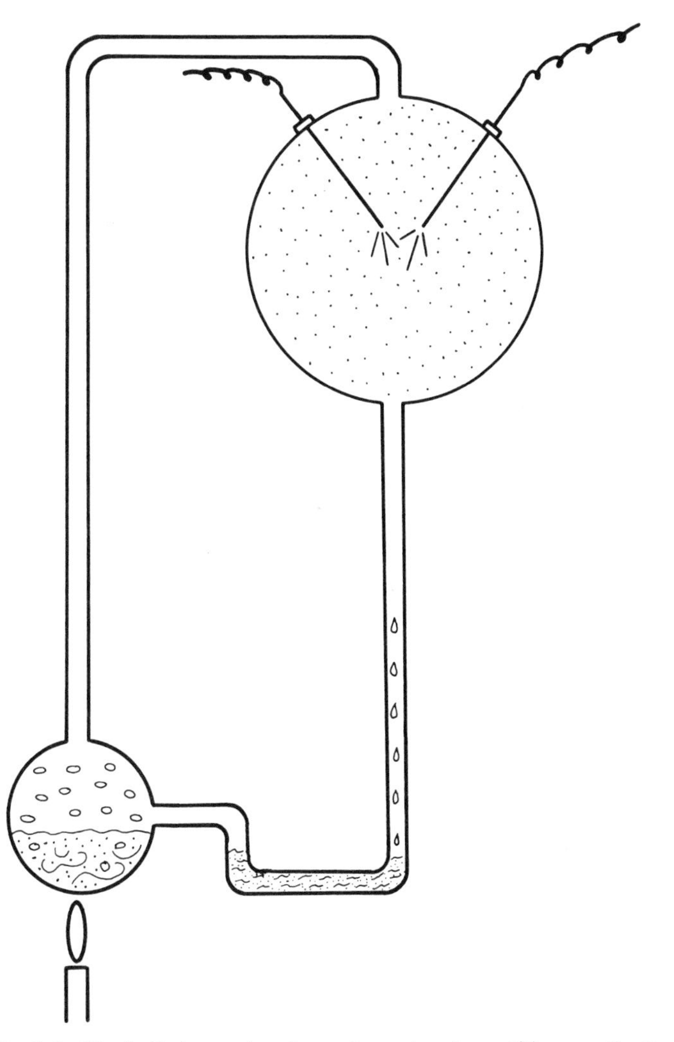

Fig. 1-8. Spark discharge chambers represent early conditions on Earth.

The carbon chains in Earth's primordial seas strung together to form molecules of hydrogen cyanide, ethane, ethylene, and formaldehyde—some of the first steps toward life. Formaldehyde molecules were particularly important because they joined to form sugars, such as *ribose,* a constituent of RNA, which is thought to be the original genetic material. RNA can evolve independently in a test tube and can replicate itself in the absence of proteins. This experiment suggests that the first organisms were simply genes.

The infant Earth of some 4 billion years ago was therefore an "RNA world"—RNA had the ability to do everything necessary for life. However, no one has yet figured out where the RNA itself came from. Its precursor molecules were probably abundant on the primitive Earth, and random chemical reactions among them might have produced a variety of molecular complexes that competed against each other. Therefore, how RNA became the dominate genetic material at this early age still remains a mystery. Perhaps through a process of spontaneous self-organization, a primitive form of metabolism operated before the arrival of RNA, thus providing a form of natural selection without the need for genetics.

A short time after the formation of the atmosphere and ocean, all the essential amino acids and nucleotides, the subunits of RNA and DNA, were present, and the ocean became a rich broth of organic compounds. The developing oceans were probably salty, although less so than modern seas. The early water bodies were diluted by comets that brought substantial amounts of water to the planet, as well as an abundance of carbon compounds.

THE INITIATION OF LIFE

Prior to the initiation of life on Earth, a form of chemical evolution was already occurring. This evolution built a stock of organic molecules, from which living organisms could be made. Possibly, in a few billion years or so, random combinations and permutations would have eventually given rise to an entity that has all the fundamental properties of life: the ability to store energy, to catalyze chemical reactions, and to self-replicate. Clay has all of these qualities and it has been around since the very beginning. Indeed, the Earth's surface can be thought of as a huge factory that manufactures clay minerals. Clays and organic molecules also coexist in meteorites.

Cycles of wetting and drying produced by the ocean tides (FIG. 1-9) cause stress in the clay that translates into energy. These cycles can link molecules of amino acids together by transferring energy from the environment to the organic molecules. The ions in clay act as catalysts to speed up chemical reactions among amino acids. When they are in the presence of clays, some organic molecules can also perform functions similar to those of enzymes.

Crystals are the most common self-assembled objects. Clay crystals replicate by spontaneously crystalizing. Seed crystals, called *crystal genes*, provide templates upon which silicon atoms and metal ions can grow, layer by layer. The crystal gene grows sideways, spreading its genetic message down lengths of folded or branched membranes or flakes. Mutation, a defect in a genetic crystal, controls the growth of interweaving crystals, possibly making the clays more suitable for their particular environment. Thus, a sort of natural selection operated in the inorganic world long before it did in the organic world.

The precursors of RNA could be used as structural materials for clays. Perhaps the negatively charged spine of RNA would stick to the edges of positively charged clay particles and "read" the information in the clay genes. It could then carry genetic information along the entire length of clay strands. A genetic takeover would occur when RNA acquired the ability to self-replicate. It could then carry the clay's genetic code elsewhere and influence the growth of other clays.

Eventually, RNA would begin to build intricate structures on its own. However, these were not made out of inorganic molecules, such as those found in

(PHOTO BY W.O. ADDICOTT, COURTESY OF USGS)

Fig. 1-9. Intertidal exposure near Pillar Point, Washington represents the early world of rock.

clay, but instead from organic molecules because they are much handier to work with. The first proteins were probably synthesized in this manner.

THE FIRST ORGANISMS

Life arose on this planet during a time of crustal formation and atmospheric and oceanic outgassing. This was also a time of heavy meteorite bombardment, which might have had a major influence on the final outcome of the planet. When proteins were first organizing into living cells, the conditions were very difficult because the early Earth was constantly being showered with comets and meteorites. Therefore, early living cells might have been repeatedly exterminated, forcing life to originate over and over again.

It is possible that whenever primitive organic molecules attempted to arrange themselves into living matter, frequent impacts blasted them apart before they could reproduce. Perhaps the only safe place for life to evolve would have been on the deep ocean floor. Hydrothermal vents (FIG. 1-10), which are like geysers on the bottom of the ocean, could have provided life with all the necessary nutrients to sustain itself. In this environment, life could have originated as early as 4.2 billion years ago.

No matter how varied life is on Earth today, from the simplest bacteria to man, its central molecular machinery is exactly the same. Every cell of every organism is constructed from the same set of twenty amino acids. All organisms use the same energy-transfer mechanism for growth. All strands of DNA are built into left-handed double helixes; no right-handed DNA strands have ever been found. Also, the operation of the genetic code in protein synthesis is the same for all living things. With so much similarity, it stands to reason that all life sprang from a common ancestor, and that any alien forms, of which there are no present-day descendents, became extinct early in the history of life on Earth.

Since life made its appearance within the first billion years of the Earth's existence, it must have evolved from simple materials into complex organisms rather quickly. Primitive bacteria, which descended from the earliest-known form of life, still remain by far the most abundant organisms. Evidence that life began very early in the Earth's history, when the planet was still quite hot, exists today in the form of *thermophilic* (heat-loving) *bacteria*, found in thermal springs and other hot-water environments throughout the world (FIG. 1-11).

Because these bacteria have no nucleus, which would cease to function in hot water, they can live and reproduce successfully, even at temperatures well above

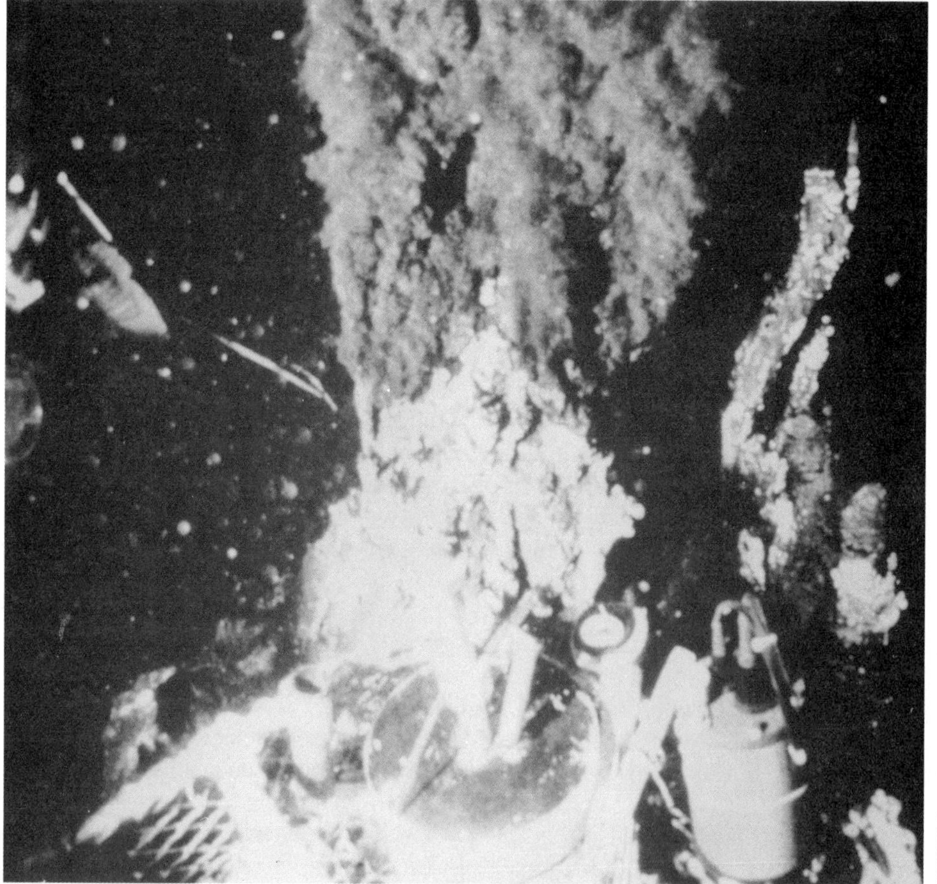

(PHOTO BY N.P. EDGAR, COURTESY OF USGS)

Fig. 1-10. Hydrothermal vent on the ocean floor at the East Pacific Rise spews black sulfide-rich hot water.

(PHOTO BY W.H. JACKSON, COURTESY OF USGS)

Fig. 1-11. Mammoth Hot Springs, Yellowstone National Park, Wyoming.

the normal water boiling point (as long as it remains a liquid). These conditions require high pressures equal to those found in the deep sea near hydrothermal vents. The existence of these organisms is often used as evidence that thermophiles were the common ancestors of all life on Earth.

The early conditions on Earth would have been ripe for the evolution of thermophilic organisms. Most thermophiles have a sulfur-based energy metabolism, and sulfur compounds would have been plentiful on the hot, volcanically active planet (FIG. 1-12). It was very fortunate that the early Earth had abundantly spewed sulfur from a profusion of volcanoes. With the absence of a solar ultraviolet shield, which today is generated by the ozone layer, the first living cells would have met a sizzling death in the deadly rays of the Sun.

As long as surface temperatures remained fairly hot, ring molecules of sulfur atoms in the atmosphere would have effectively blocked the ultraviolet radiation. However, an ultraviolet shield might not have been necessary in the primordial atmosphere, because some primitive bacteria appear to be able to tolerate high levels of ultraviolet radiation.

The first living organisms were extremely small noncellular blobs of protoplasm, called *prokaryotes*, which did not have an organized nucleus of genetic material. These organisms reproduced asexually by simple fission, whereby the organism either split in two or small parts of it budded off and began growing independently. This form of reproduction provided little or no variation among species. Plenty of food was available, and the self-duplicating organisms lived on an abundance of organic molecules in the primordial sea.

These organisms must have set off a rapid chain reaction that resulted in phenomenal growth. The organisms drifted freely in the ocean currents until they were dispersed to all parts of the world. Although the first simple organisms appear to have arrived fairly soon after conditions on Earth became favorable, it took almost another billion years before life would remotely resemble what it is today.

THE FIRST EXTINCTION

Since life's first humble beginnings, it has responded to a variety of chemical, climatological, and geographical changes in the Earth, forcing species to either adapt or perish. Throughout geologic history, vast numbers of species have vanished. During geologically brief intervals of several million years, mass extinctions have eliminated most of the species of plants and animals on Earth.

Devastation of this magnitude could only have been inflicted by radical changes in the environment on a global scale. A number of theories have been developed to explain the cause of this phenomenon. In one theory, cosmic radiation from a nearby supernova or from a massive meteorite bombardment (FIG. 1-13) scorched the Earth. There might also have been drastic changes in the environmental limiting factors that included the temperature and living space on the ocean floor, both of which determine the distribution and abundance of species in the sea.

(PHOTO BY C.W. STOUGHTON, COURTESY OF NATIONAL PARK SERVICE)

Fig. 1-12. Haleakala Crater on Hawaii represents conditions that existed on the early Earth.

Fig. 1-13. The heavily cratered Moon indicates that the Earth was equally bombarded by meteorites in its early stages. (PHOTO BY H.A. POHN, COURTESY OF USGS)

The most important factor limiting the geographical distribution of marine species is water temperature. Some species can only survive within a narrow temperature range. An episode of climatic cooling could extinguish any species that could not adapt to the new, colder temperature or migrate to a warmer refuge. One such event might have occurred 2 billion years ago, when Earth plunged into an early ice age that might have eliminated primitive life forms that were attempting to evolve during this time.

The first major mass extinction in the fossil record occurred in the late Precambrian era, about 670 million years ago. At that time, animal life was still very sparse, and this extinction decimated the ocean's population of single-celled phytoplankton, the first organisms to have cells with nuclei. The mass disappearance of this species coincided with the period when glaciers covered much of the Earth's surface. When the ice disappeared, development of new species exploded and forever changed the composition of life on Earth.

2

The History of Life

THE history of the Earth is written in its rocks, and the history of life is told by its fossils. Many evolutionary paths weave their way along branches of the tree of life, and species leave their footprints in the fossil record, which itself only represents a fraction of those that have ever lived. Most every conceivable form and function have been tried, some more successfully than others. It is through this trial and error method of specialization that by natural selection certain species prosper and others become biological dead ends.

AGE OF BEGINNING LIFE

The first 4 billion years, or about 90 percent of geologic time (TABLE 2-1), constitutes the *Precambrian era*, the longest and least-understood period of Earth history. The Precambrian is divided nearly in half by the *Archean* and *Proterozoic eons*. During the Archean eon, which spans from 4.6 to 2.5 billion years ago, the Earth was in great turmoil and was subjected to extensive volcanism and meteorite bombardment. These activities probably had a major effect on the creation of life so early in the planet's history.

Archean life consisted mostly of bacteria and primitive algae. These organisms lacked a distinct nucleus and are called *prokaryotes*, derived from the Greek word *karyo*, meaning nut. They lived in anaerobic conditions with an absence of oxygen. The organisms depended mostly on outside sources of nutrients, which

TABLE 2-1. The Geologic Time Scale

ERA	PERIOD	EPOCH	AGE IN MILLIONS OF YEARS	FIRST LIFE FORMS
		Holocene	0.01	
	Quaternary			
		Pleistocene	2	Man
Cenozoic		Pliocene	7	Mastodons
		Miocene	26	Saber-tooth tigers
	Tertiary	Oligocene	37	
		Eocene	54	Whales
		Paleocene	65	Horses Alligators
	Cretaceous		135	Birds
Mesozoic	Jurassic		190	Mammals
	Triassic		240	
	Permian		280	Reptiles
		Pennsylvanian	310	
	Carboniferous			Trees
Paleozoic		Mississippian	345	Amphibians Insects
	Devonian		400	Sharks
	Silurian		435	Land plants
	Ordovician		500	Fish
	Cambrian		570	Sea plants Shelled animals
			700	Invertebrates
Proterozoic			2500	Metazoans
			3500	Earliest life
Archean			4000	Oldest rocks
			4600	Meteorites

consisted of a rich supply of organic molecules that were constantly being generated in the sea around them.

Photosynthesis as an energy source might have begun as early as 3.5 billion years ago. The first organisms to use it as their main source of energy were simple bacteria. Like those living today, early bacteria were best suited to an oxygen-poor environment. Therefore, oxygen, which is poisonous to primitive life-forms, was reduced by chemical reactions with dissolved metals in the seawater.

The oldest evidence of life on Earth are *microfossils*, the remains of ancient microorganisms, and *stromatolites* (FIG. 2-1), the layered structures formed by the accretion of fine sediment grains by colonies of primitive bacteria. The earliest stromatolites were found in 3.5-billion-year-old sedimentary rocks of the Warrawoona group in Western Australia. Associated with these rocks were cherts that contained *microfilaments*, small, threadlike structures of possible bacterial origin. Similar cherts with microfossils of filamentous bacteria were found in 3.3-billion-year-old rocks from eastern Transvaal in South Africa.

(PHOTO BY R. REZAK, COURTESY OF USGS)

Fig. 2-1. Stromatolite of the Missoula group on Hidden Lake trail, Glacier National Park, Montana.

The abundance of chert in deposits older than 2.5 billion years indicates that most of the crust was deeply submerged at this time. The seas contained much more dissolved silica, which leached out of volcanic rock that poured onto the ocean floor. Modern ocean water is deficient in silica because organisms like sponges and diatoms extract it to build their skeletons. Massive deposits of diatomeceous earth are a tribute to the success of these organisms in the post-Precambrian era.

Stromatolite structures are indirect evidence of life because they are not the remains of the microorganisms themselves, but only the sedimentary structures they built. The Australian stromatolites are distinctly layered accumulations of calcium carbonate with a rounded, cabbagelike appearance. Modern stromatolites are very similar, composed of concentrically layered mounds of calcium carbonate built by bacteria or algae. These organisms cement sediment grains by secreting a jellylike ooze from their cells, which binds the sediment into thin laminate.

The organisms photosynthesized for energy. Therefore, they were dependent on sunlight, so they had to live near the water's surface. Just as modern organisms do today, Precambrian stromatolite colonies lived in the intertidal zones. Their great height, upwards of 30 feet or more, was indicative of the tide height. The tides rose to tremendous heights because the Moon was only about half its present distance away. Therefore, it pulled much harder on the oceans, piling water up into a huge bulge that followed the Moon around the Earth.

The sinuous growth pattern in the ancient stromatolites also indicates the number of days in the year. About a billion years ago, a year contained nearly 450 days; Earth was spinning much faster on its axis, completing a single day in about 20 hours.

AGE OF EARLY LIFE

The Proterozoic eon, from 2.5 to 0.6 billion years ago, represented a dramatic change in the Earth as it matured from a tumultuous childhood to a much calmer adulthood. When the eon began, as much as 75 percent of the current continental crust had formed. Continents stabilized and gathered into a single large supercontinent near the equator. Marine life of the Proterozoic was much more distinctive than that of the preceding Archean and it represented a considerable biological advancement.

Proterozoic life also showed much more complexity than Archean life. Organisms appear to have evolved very little during their first billion-year-stay on Earth. This process might have been slow as a result of their primitive form of reproduction, involving simple fission. Thus, fewer mutations occurred, providing little or no genetic change. Moreover, a primitive form of metabolism that utilized fermentation required organisms to live in a low-energy state.

The first major advancement in the Proterozoic came with the development of an organized nucleus and sexual reproduction around 1.4 billion years ago. This advancement resulted in a new breed of single-cell organisms, called *eukaryotes.* Apparently with this development, evolution suddenly speeded up. Metabolism in eukaryotes was accomplished by respiration. Therefore, the presence of these organisms indicates that the atmosphere and ocean contained substantial amounts of oxygen at this time (TABLE 2-2).

TABLE 2-2. Evolution of Life on Earth

EVOLUTION	ORIGIN (MLLION YEARS)	ATMOSPHERE
Origin of Earth	4600	Hydrogen, Helium
Origin of life	3800	Nitrogen, methane, carbon dioxide
Photosynthesis	2300	Nitrogen, carbon dioxide, oxygen
Eukaryotic cells	1400	Nitrogen, carbon dioxide, oxygen
Sexual reproduction	1100	Nitrogen, oxygen, carbon dioxide
Metazoans	700	Nitrogen, oxygen
Land plants	400	Nitrogen, oxygen
Land animals	350	Nitrogen, oxygen
Mammals	200	Nitrogen, oxygen
Man	2	Nitrogen, oxygen

The first single-celled animals, called *protistids,* shared many characteristics with plants, except that they could move under their own power. Some animals traveled by thrashing their whiplike tails, called *flagella.* Other animals had tiny hairlike appendages, called *cilia,* which helped them move by rhythmically beating the water. The amoeba traveled along the ocean floor by extending fingerlike protrusions outward from its main body and flowing into them. This mobility enabled animals to feed on plants and other animals, thus establishing the predator-prey relationship that became one of the most important factors in the diversification of species.

Multicellular animals, called *metazoans,* evolved in the latter part of the Proterozoic, around 700 million years ago, when the oxygen in the ocean reached nearly 10 percent of its present level. The first metazoans were a loose organization of individual cells united for common purposes, such as locomotion, feeding, and protection. The most primitive metazoans were probably composed of a large number of cells, each with their own flagellum. The cells were grouped into a small, hollow sphere, and their flagella beat the water in unison to propel the tiny animal.

Other metazoans evolved into sedentary types that turned inside out and attached to the ocean floor. They had openings to the outside and the flagella, now on the inside, produced a flow of water, providing a crude circulatory system for filtering food particles and ejecting waste products. These were the forerunners of the sponges (FIG. 2-2). Some species grew as large as ten feet or more across and were the first giants of the sea.

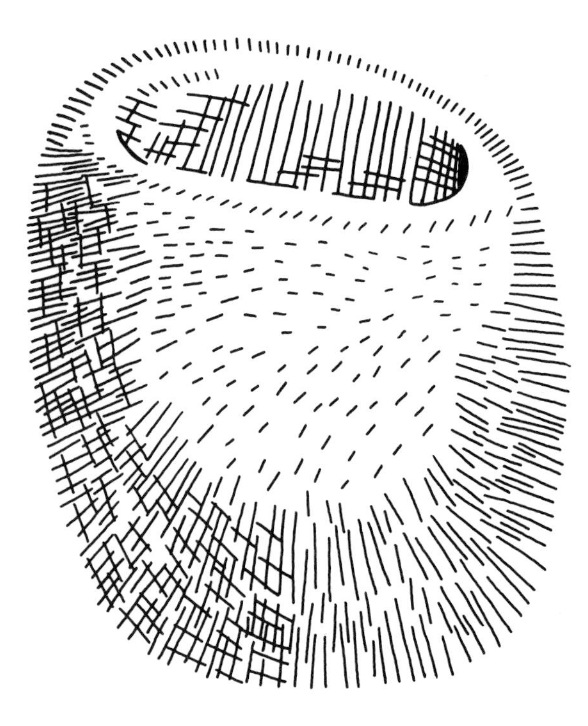

Fig. 2-2. Early sponges were among the first giants on the ocean floor.

On the next step were the jellyfish, which had two layers of cells separated by a gelatinous substance. This structure provided the animal with a means of support. Unlike the sponges, however, the cells of the jellyfish were incapable of independent survival after becoming detached from the main body. The cells were controlled by a primitive nervous system, thus becoming the first life-form with simple muscles for locomotion.

The development of muscles and other rudimentary organs, including sense organs and a central nervous system, came with the evolution of the primitive segmented worms. They left behind a preponderance of fossilized tracks, trails, and burrows; so much so that the Proterozoic has often been described as the "age of worms." Prior to 670 million years ago, however, there were no track-making animals.

By the time the Proterozoic closed, the sea contained large populations of widespread and diverse creatures. The dominant organisms were the *coelenterates*, including giant jellyfishlike floaters that grew up to three-feet wide. Also colonial feathery forms, probably ancestors of the corals, attached to the seafloor and grew more than a yard long. The remaining creatures were mostly marine worms, unusual naked arthropodlike animals, and tiny curious-looking "naked" starfish that had three rays instead of the customary five.

Many strange species were found in the Ediacara formation of southern Australia from around 670 million years ago. Many of these unusual creatures (FIG. 2-3) probably resulted from adaptations to the highly unstable conditions that existed in the late Proterozoic, along with an increasing oxygen supply. As a consequence of over-specialization, a major extinction of species occurred at the end of the era around 570 million years ago. Those species that survived the great extinction were quite different from their Ediacarian ancestors. These new life-forms flourished in the warm Cambrian seas and participated in the greatest explosion of new species in Earth history.

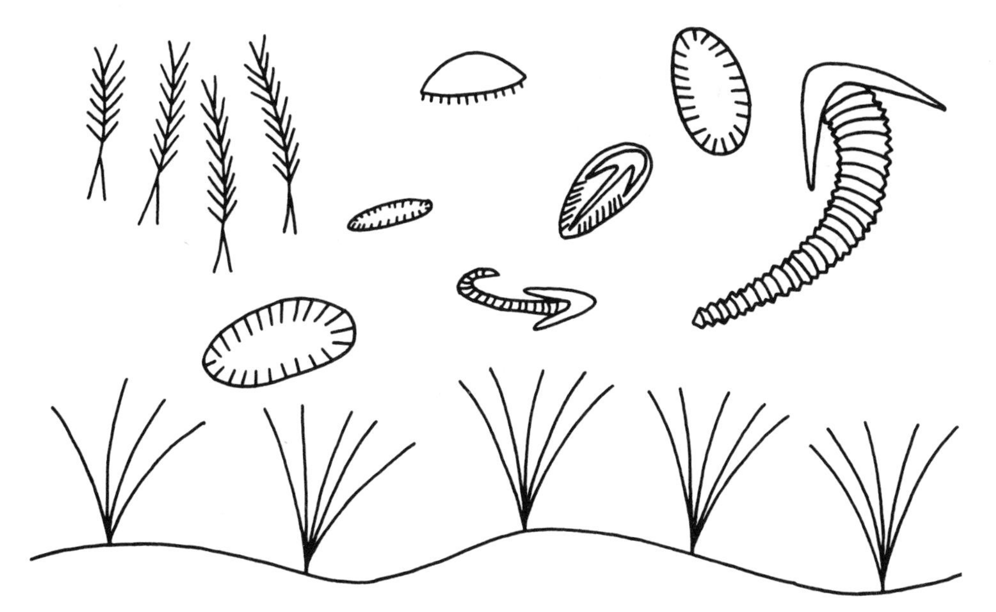

Fig. 2-3. The Precambrian Ediacara fauna found in the Ediacara Hills of South Australia.

AGE OF ANCIENT LIFE

The Paleozoic era, "the age of fishes," spans from about 570 million to about 240 million years ago. This time of intense growth and competition produced widely dispersed and diversified species (FIG. 2-4). By the middle of the era, all major animal and plant *phyla*—groups of organisms that share the same general body plan—were already in existence. When the era came to an end, however, the greatest mass extinction Earth has ever known left it almost as devoid of life as when the era began.

The Cambrian is best known as the "age of the trilobites" (FIG. 2-5). They appeared at the very beginning of the period and became the dominant species of

Fig. 2-4. Early Paleozoic fossils on display at the Museum of Geology, South Dakota School of Mines at Rapid City.

the early Paleozoic. Successfully sharing the sea bottom with the trilobites were the *brachiopods* or lamp shells, which had two clamlike shells fitted face-to-face that opened and closed with simple muscles. The brachiopods were fixed to the ocean floor by a sort of rootlike appendage. They fed themselves by filtering food particles through their opened shells.

The crinoids, also known as sea lilies (FIG. 2-6), were another dominant species of the middle and late Paleozoic and are still living today. They have long stalks, sometimes over ten feet long, composed of hundreds of calcite disks anchored to the seafloor with a rootlike structure. A cup, called a *calyx*, perches on top of the stalk, housing the digestive and reproductive systems. *Blastoids*, similar to crinoids, became extinct at the end of the Paleozoic as a result of stiff competition from their crinoid cousins for scarce shallow-water platform habitats.

The prolific sponges of the early Paleozoic were one of the most primitive of all species. They came in a variety of shapes and sizes and grew in thickets on the sea floor. Competing along with the sponges were the corals (FIG. 2-7), which had hard skeletons and successive generations built thick limestone reefs. The corals began constructing reefs in the early Paleozoic, they formed entire chains of islands and altered the shorelines of the continents. Many corals became extinct and were replaced by sponges and algae in the late Paleozoic as a result of the retraction of the seas in which they once thrived.

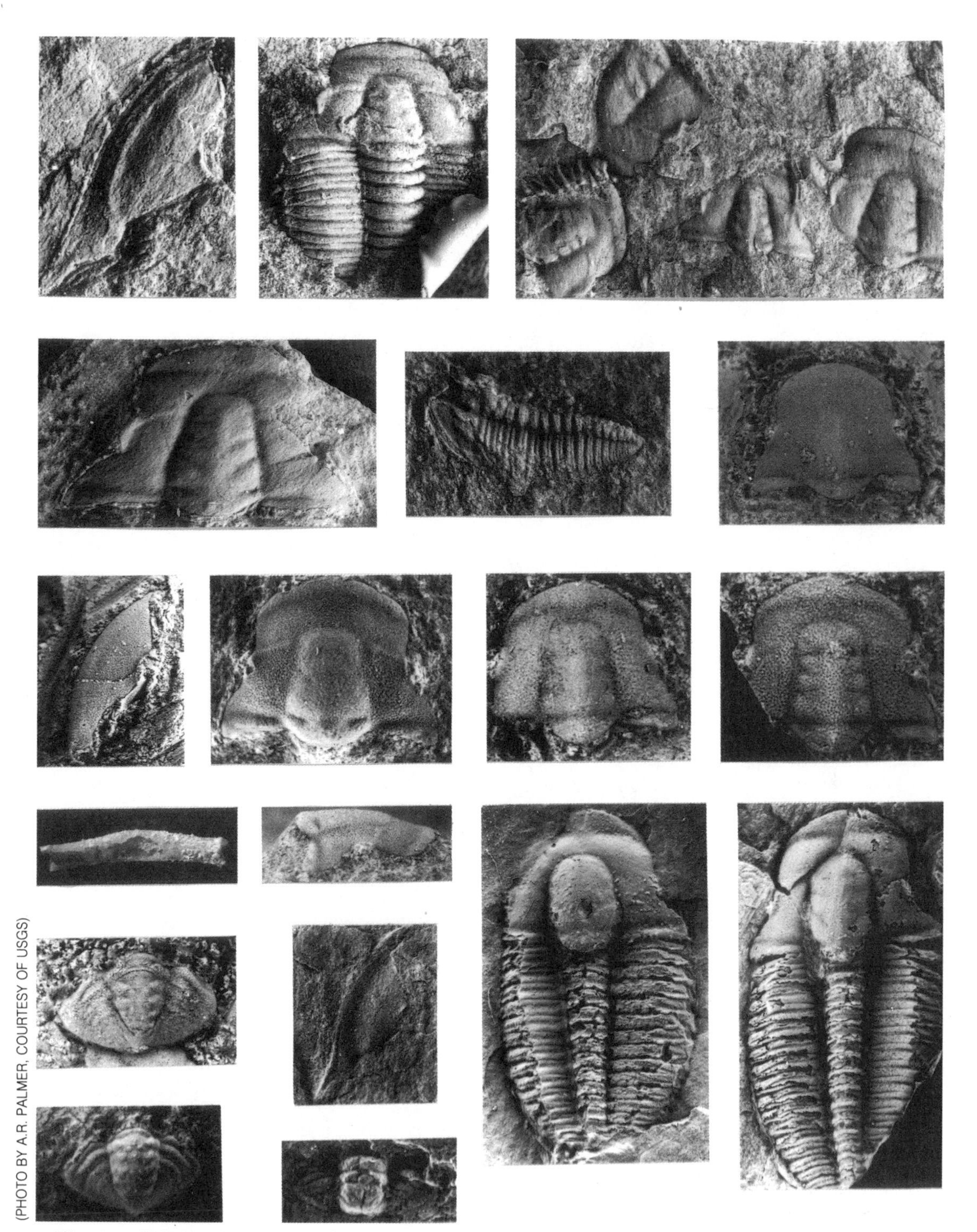

(PHOTO BY A.R. PALMER, COURTESY OF USGS)

Fig. 2-5. Trilobite fossils of the Carrara Formation in the Southern Great Basin, California/Nevada.

Fish achieved dominance in the Devonian period, and the fossil record reveals so many varied kinds of fish that *paleontologists* (geologists who study fossils) have a difficult time classifying them all. Ancient fish were probably poor swimmers and avoided deep water. Bony plates surrounded the head for protection, and the rest of the body was covered with thin scales. Jawless fish (FIG. 2-8) are considered the earliest known vertebrates, having been in existence for 470 million years. Freshwater fish living in Australia around 370 million years ago were almost identical to those living in China. This example suggests that the two landmasses were close to each other at this time, allowing the fish to travel between them.

Like the animals, plants did not appear in the fossil record as complex organisms until the Cambrian, after which they began to evolve rapidly. During the Ordovician, around 450 million years ago, the oxygen level was sufficient to create an effective ozone screen in the upper stratosphere. This layer shielded the Earth's surface from the Sun's deadly ultraviolet rays, enabling plants and animals to come ashore for the first time.

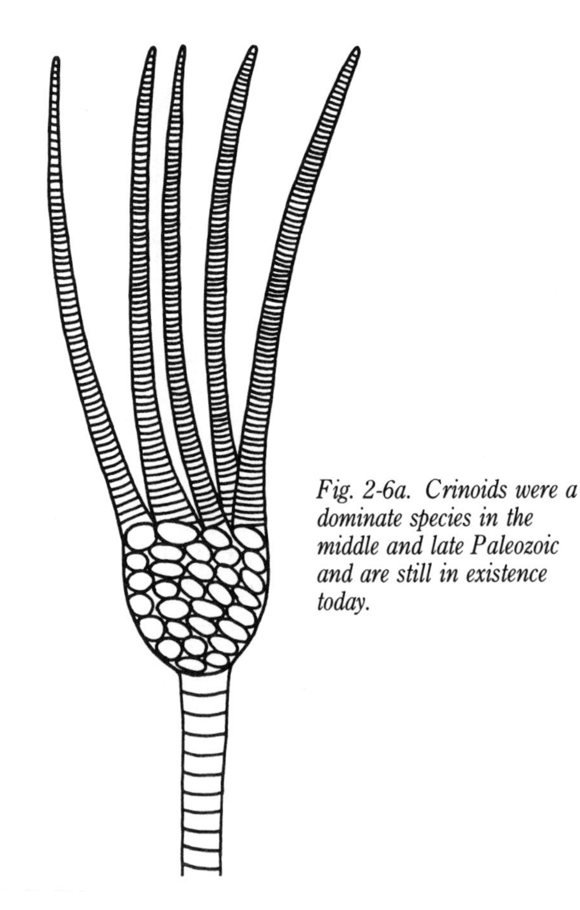

Fig. 2-6a. Crinoids were a dominate species in the middle and late Paleozoic and are still in existence today.

(PHOTO BY R.C. MCDOWELL, COURTESY OF USGS)

Fig. 2-6b. Crinoid columnals in the so-called "bead bed" of the Drowning Creek Formation in Fleming County, Kentucky.

When the early plants first left the water for a home on dry land, they were greeted by a harsh environment, where drought, ultraviolet radiation, and lack of nutritional sources made life exceedingly difficult. In order to survive these rugged conditions, plants had to rely on symbiotic relationships, whereby one organism lives off the waste products of the other. *Lichens*, a partnership between algae and fungus, probably took the first tentative steps on land. They were followed by the mosses and liverworts. The early land plants diverged into two major groups. One gave rise to the *lycopods*, or club mosses, and the other gave rise to the *gymnosperms*, which were ancestral to the great majority of modern land plants.

The first invertebrates to crawl out of the sea were probably crustaceans. These segmented creatures, ancestors of today's millipedes, walked on perhaps a hundred pairs of legs. They also became easy prey when the descendents of the giant sea scorpion eurypterid (FIG. 2-9) began to come ashore. By the middle Devonian, about 370 million years ago, stiff competition in the sea encouraged crossopterygians and their relatives to make short forays on shore to dine on abundant crustaceans and insects. By late Devonian, these fish probably gave rise to the earliest amphibians.

(PHOTO BY P.E. CLOUD, COURTESY OF USGS)

Fig. 2-7. A collection of corals off the coast of Saipan in the Marianas Islands.

Fig. 2-8. The jawless fish, which lived 470 million years ago, were the earliest vertebrates.

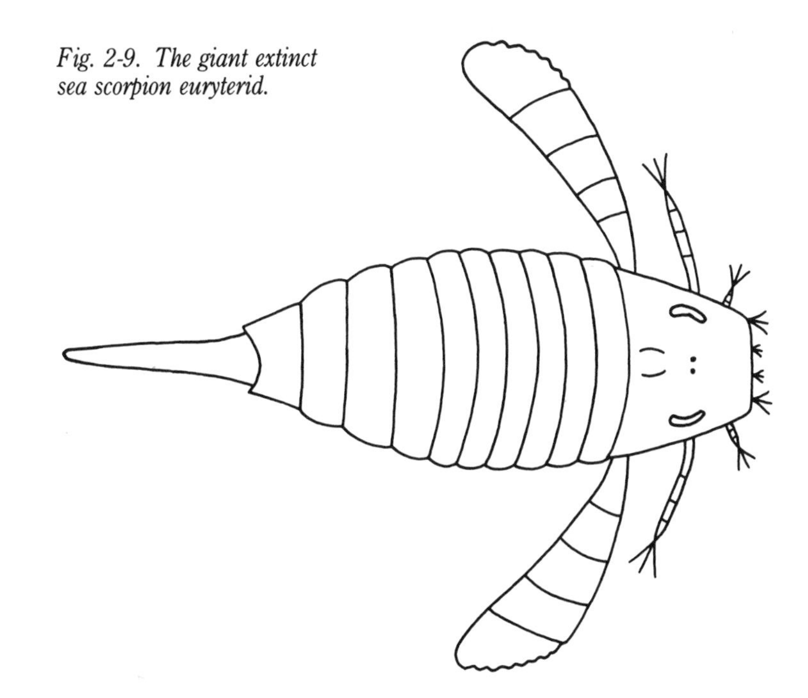

Fig. 2-9. The giant extinct sea scorpion euryterid.

Amphibians live a semiaquatic life-style, which led to their eventual downfall when the great swamps dried up toward the end of the Paleozoic. The space left by the amphibians was quickly filled by their cousins, the reptiles, which were much better adapted to a life out of water. The reptiles probably eventually gave rise to the dinosaurs, the most successful species the world has ever known.

AGE OF MIDDLE LIFE

The *Mesozoic era*, "the age of dinosaurs," spans from about 240 to about 65 million years ago (FIG. 2-10). When the era began, the Earth was recovering from a major ice age and the worst extinction event in geologic history. Those species that were immobile and could not migrate to better habitat or had developed specialized life-styles and were unable to adapt to a changing environment were the hardest hit. Moreover, those species that survived the great extinction differed markedly from their close relatives that were left behind.

Thus, the beginning of the Mesozoic was a sort of rebirth of life, and 450 new families of organisms came into existence. However, instead of inventing entirely new body plans, as was the case with the Cambrian explosion, the beginning of the Mesozoic saw only new variations on already established themes. Therefore, there were much fewer experimental organisms, but many lines of modern species began at this time.

Fig. 2-10. The major Cretaceous dinosaurs.

The Mesozoic was also a period of transition, especially for the plants, which at the beginning of the era showed little resemblance to those living today (FIG. 2-11). A radical change in vegetation occurred during the middle Cretaceous with the introduction of the *angiosperms* (flowering plants), which began alongside pollinating insects. By lowering the level of atmospheric carbon dioxide, which lowered global temperatures, the rise of the angiosperms also might have contributed to the extinction of the dinosaurs and certain marine species at the end of the Cretaceous.

In the latter part of the Paleozoic, the reptiles largely replaced the amphibians and became the dominant land-dwelling animals of the Mesozoic. The generally warm climate of the Mesozoic worked to the reptile's advantage and greatly assisted them with colonizing the land. The most successful reptiles were the dinosaurs, which occupied a wide variety of ecological niches and dominated all other forms of land-dwelling animals. The oldest dinosaurs stemmed from Gondwana following the great Permian ice age. These thick glaciers covered much of the southern continents, which were locked together over the South Pole.

Birds descended from the *thecodonts*, which were the same ancestors of the dinosaurs. Therefore, birds are often referred to as "glorified reptiles." They are warm-blooded in order to obtain the maximum metabolic efficiency needed for sustained flight. However, they retain the reptilian mode of reproduction by laying eggs. The bird's ability to maintain its body temperatures has led to speculation that some dinosaur species were also warm-blooded.

Fig. 2-11. A fossil leaf from the Raton Formation near Trinidad, Colorado shows many features of modern leaves. (PHOTO BY W.T. LEE, COURTESY OF USGS)

The first animals to diversify from the reptiles were the *pelycosaurs* (FIG. 2-12), which began about 300 million years ago. They were distinguished by their large size; they obtained a length of about eleven feet. They had large dorsal sails composed of webs of membrane stretched across bony, protruding spines. The sails were well-supplied with blood and probably used to regulate the animal's body temperature. The pelycosaurs thrived for about 50 million years and then gave way to their descendents, the mammal-like reptiles, called *therapsids*, which ranged from mouse-sized to hippopotamus-sized.

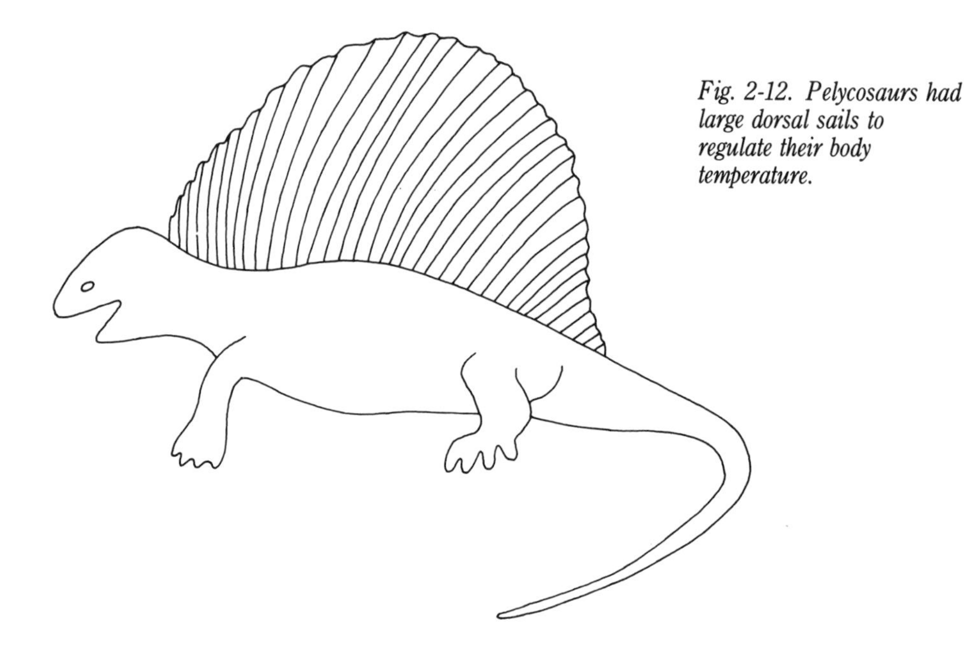

Fig. 2-12. Pelycosaurs had large dorsal sails to regulate their body temperature.

The family of mammal-like reptiles clearly shows a transition from reptiles to mammals. The early members invaded the southern continents during the late Permian when those lands were still quite cold, which suggests that the therapsids might have been warm-blooded. The development of fur appeared in the more advanced therapsids that moved into the colder climates. The therapsids dominated animal life on Earth for more than 40 million years until the middle of the Triassic period. Then for unknown reasons, they lost out to the dinosaurs.

The boundary between the end of the Triassic period and the beginning of the Jurassic period, around 210 million years ago, was one of the most exciting times in the history of land vertebrates. By then, almost all modern animal groups, including amphibians, reptiles, and mammals, had made their debut. This time was also when the dinosaurs achieved dominance over the Earth and held their ground for the next 140 million years.

(PHOTO BY J.P. IDDINGS, COURTESY OF USGS)

Fig. 2-13. Fossil tree trunks at Specimen Ridge, Yellowstone National Park, Wyoming.

AGE OF RECENT LIFE

The Cenozoic era, "the age of the mammals," covers the last 65 million years of Earth history. Extremes in climate and topography resulted in a greater variety of living conditions than during any other equivalent span of geologic time. The rigorous environments presented many challenging opportunities for species, and the extent to which plants and animals invaded diverse habitats was truly remarkable. Mammals began to rapidly diversify by the end of the Paleocene epoch, about 54 million years ago. At the same time, some large, peculiar-looking animals became extinct. By the start of the Eocene epoch, most of the truly modern mammals began to appear.

Marine species that made it through the great extinction at the end of the Cretaceous, which took the dinosaurs and large numbers of other species, looked much the same as they did in the Mesozoic. Species that inhabited unstable environments, such as those found in the higher latitudes, were especially successful. Some 70 species of marine mammals, known as *Cetaceans,* which include dolphins, porpoises, and whales appeared during the middle Cenozoic.

All major groups of modern plants were represented in the early Cenozoic. The angiosperms dominated the plant world, and all modern families appear to have evolved by the Miocene epoch (about 25 million years ago). Grasses were the most important angiosperms, providing food for hoofed mammals, called *ungulates,* throughout the Cenozoic. Many large mammals probably began grazing in response to the widespread availability of grassland.

At one time, forests of giant hardwood trees grew as far North as Yellowstone National Park (FIG. 2-13). Today, only scraggly conifers grow on this land, indicating that the climate is presently cooler. The cone-bearing plants, which were prominent during the Mesozoic, occupied a secondary role during the Cenozoic. Tropical vegetation, which was widespread during the Mesozoic, withdrew to narrow regions around the equator in response to a colder, drier climate. These climatic conditions were mainly caused by a general uplift of the landmasses, which resulted in the draining of the interior seas.

Most scientists believe that primates evolved from a small squirellike mammal that lived more than 50 million years ago. Afterwards, the primate family tree split into two branches: the New World monkeys and the Old World monkeys. The latter group also included the hominoids, our humanlike ancestors. Approximately 30 million years ago, the precursors of the apes lived most of their lives in the dense tropical rain forests of Egypt, which today is mostly desert. These apelike ancestors spread from Africa into Europe and Asia between about 25 million and 10 million years ago.

The common ancestor of apes and humans is thought to be *Proconsul,* an apelike creature that lived from about 22 million to 16 million years ago. It was a tree-dwelling, fruit-eating animal that walked mainly on all fours. Also, males were distinctly larger than females. Proconsul evolved after monkeys became a separate branch of the primate family tree, but before chimpanzees, gorillas, orangutans, and humans branched off. Little is known about human beginnings, from about 14 million to four million years ago, when the human line apparently split away.

3

The Major Extinctions

THROUGHOUT the history of life on Earth, species have come and gone; those living today only represent a tiny fraction of the total. More than 99 percent of all species that have inhabited the Earth at one time or another are extinct. It is estimated that as many as four billion species of plants and animals once existed in the geologic past. Most of these lived during the Phanerozoic eon (the last 570 million years). This period featured a phenomenal development of species, along with tragic episodes of mass extinctions. Each extinction involved the disappearance of more than 50 percent of the species that lived during that time. Thus, the extinction of species has been almost as common as the origination of species.

During the great Cambrian explosion, which began about 570 million years ago (mya), species diversity was the greatest it had ever been on Earth. Since then, five major mass extinctions have occurred: at the end of the Ordovician (440 mya), the Devonian (370 mya), the Permian (240 mya), the Triassic (210 mya), and the Cretaceous (65 mya) periods. Five or more lesser mass extinctions have also occurred. The common thread running through all mass extinctions is that the biological systems were in extreme stress, mainly brought on by rapid and extensive changes in the environment.

EARLY EXTINCTIONS

The development of photosynthesis was possibly the single most-important feature in the evolution of life on Earth. It provided a primitive form of blue-green algae with a practically unlimited source of energy. Photosynthesis utilizes carbon dioxide and sunlight to manufacture carbon compounds, liberating oxygen as a byproduct. However, oxygen is also poisonous to all organisms that have not developed special defenses against it. Thus, if oxygen had continued to build up in the ocean, early life on Earth certainly would have been in jeopardy.

During the early Proterozoic, two billion years ago, when the oxygen level in the ocean and atmosphere was still only about one percent of its present value, organisms developed a new method of obtaining energy by combining oxygen with nutrients—*respiration.* This method freed many organisms from dependance on sunlight, and as a result, the first simple animals evolved. At first, little distinction existed between plants and animals, and they shared similar characteristics, utilizing both photosynthesis and respiration. As animals became more mobile and complex, they began to rely totally on respiration for their energy supply.

With carbon dioxide and other greenhouse gases depleted in favor of oxygen, the Earth cooled dramatically, bringing on a major ice age about 2 billion years ago. The colder climate probably forced many species extinct and kept the population growth down. Cold water temperatures restrict the geographic distribution of animal species and confine populations to regions that surround the equator.

Cold ocean temperatures also slow chemical reactions, as well as biological activities, thus forcing organisms to live at a lowered level of growth. When an ice age ends, generally biological diversity spurts and entirely new arrays of species come into existence. During the late Precambrian ice age, around 670 million years ago, glaciers covered more than 50 percent of the land surface and instigated the first major mass extinction. When the ice age ended, the population grew phenomenally and that diversification of species has never been equaled.

Much of the life in the ocean today has had ancestors in the late Precambrian. However, large extinctions that occurred during the Phanerozoic eliminated most species. The only records of their existence are found in fossils, which themselves only represent a fraction of all life that has ever lived on Earth. After the development of metazoans about 1.4 billion years ago, the previous skimpy fossil record began to grow. The fossil record also improved greatly 700 million years later with the advent of tracks and body imprints made by soft-bodied, burrowing animals.

When the Precambrian closed, new species exploded and for the first time, hard external skeletons provided an excellent fossil record of animal existences. The hard outer body surface also restricted the absorption of oxygen directly from seawater by exposed body surfaces. This action prompted the development of gills and circulatory systems when the oxygen level in the ocean rose to about 10 percent of its present value (FIG. 3-1).

ANCIENT EXTINCTIONS

The earliest vertebrates were the jawless fish, which lived about 470 million years ago. Fish compose over 50 percent of the species of vertebrates, both living

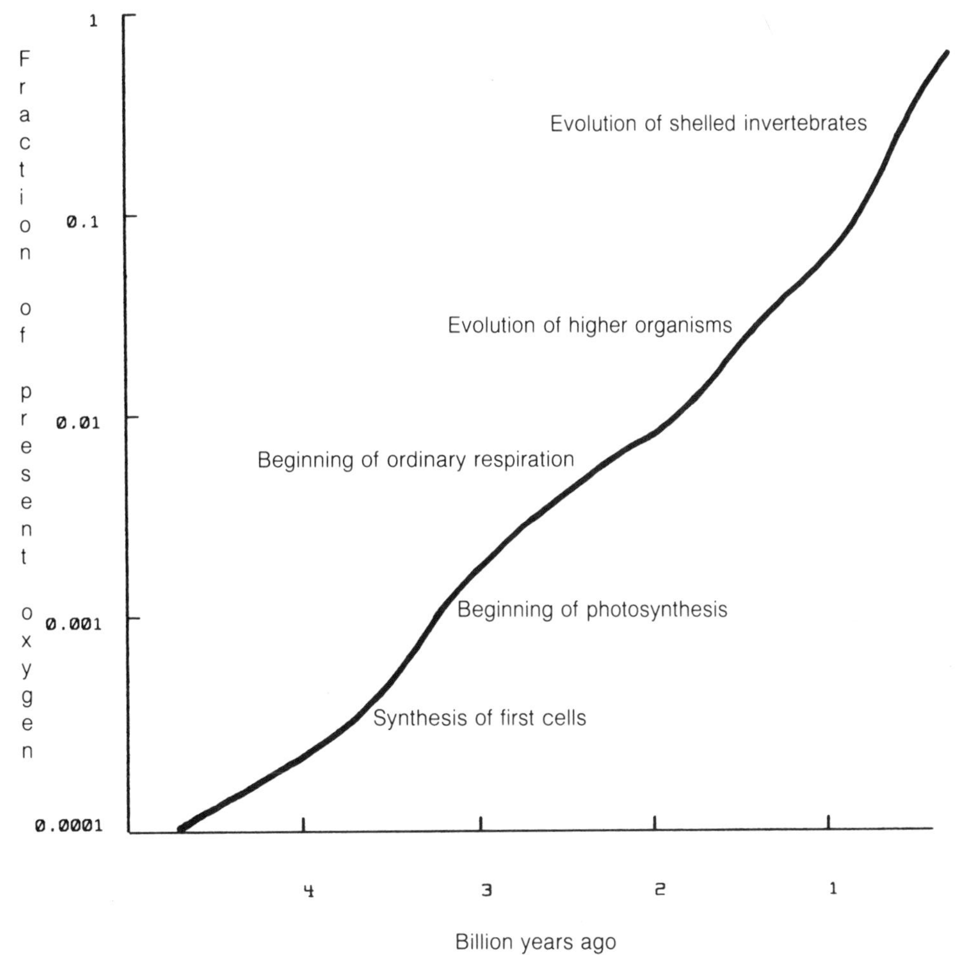

Fig. 3-1. Evolution of life and oxygen in the atmosphere.

and extinct. They progressed from having rough scales, asymmetrical tails, and cartilage in their skeletons to having flexible scales, powerful advanced fins and tails, and all-bone skeletons, much like fish today. The extinct placoderms (FIG. 3-2) were ferocious giants that reached 30 feet and more in length. The rise of fish in the Devonian seas resulted in the demise of their less mobile invertebrate competitors and largely contributed to the mass extinction at the end of the period.

The coelacanths (FIG. 3-3) were thought to have gone extinct with the dinosaurs 65 million years ago. However, in 1938, one was caught in the deep, cold waters of the Indian Ocean near Madagascar. The fish looked ancient, as though it was a castaway from the distant past. It had a fleshy tail, a large set of forward fins behind the gills, powerful jaws, and heavily armored scales. Remarkably, this fish had not changed much from its ancient ancestors, which began in the

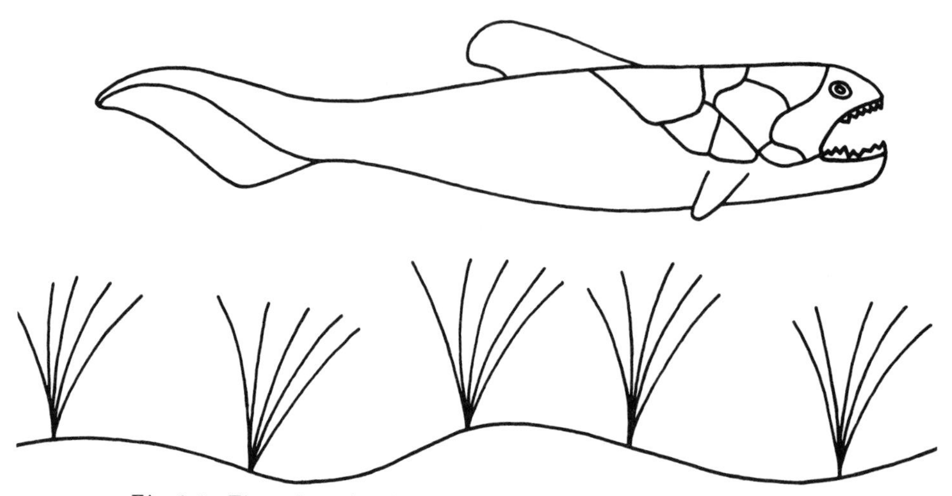

Fig. 3-2. The extinct placoderms were giants measurıng 30 feet in length.

Devonian seas 400 million years earlier. This gave coelacanth the dubious title of "living fossil."

The greatest extinction on Earth occurred at the end of the Permian period, 240 million years ago. Half the families of marine organisms, involving over 95 percent of all known species, abruptly disappeared. The temperature of the ocean was probably still low as a result of the late Permian ice age. Those marine

Fig. 3-3. Modern coelacanths have not changed much from their Paleozoic ancestors.

(PHOTO BY M. GORDON, JR., COURTESY OF USGS)

Fig. 3-4. A variety of fossil ammonite shells.

invertebrates that managed to escape extinction were forced to live in a narrow margin near the equator. Corals, which require warm shallow water, were particularly hard hit, as is evidenced by the lack of coral reefs in the early part of the Triassic period.

The crinoids and brachiopods, which had their heyday in the Paleozoic era, were relegated to minor roles during the Mesozoic era. The trilobites, which were extremely successful during the Paleozoic, died out at the close of the era. The space vacated by the trilobites was occupied by a variety of other crustaceans, such as shrimps, crabs, crayfish, and lobsters. The sharks regained ground lost during the great extinction and went on to become the successful predators they are today.

The molluscs appeared to have weathered the hard times of the late Permian extinction quite well and continued to become the most important shelled invertebrates of the Mesozoic seas, with 60,000 separate species living today. The warm climate of the Mesozoic influenced the growth of giant species in the ocean as well as on land. Giant clams grew up to three feet long, giant squids up to 65 feet long (weighing over a ton), and crinoids up to 60 feet long.

The cephalopods became the most spectacular, diversified, and successful marine invertebrates of the Mesozoic. The coiled-shelled ammonites (FIG. 3-4), which began in the early Devonian period (395 million years ago), grew as large as seven feet across. They relied on neutral buoyancy and jet propulsion for transportation, which contributed to their great success. Unfortunately, after making it through the critical transition from the Permian to the Triassic and recovering from serious setbacks in the Mesozoic, the ammonites suffered final extinction at the close of the Cretaceous period, when sea recession reduced their shallow-water habitats worldwide.

During the Cretaceous, many reptilian species returned to the sea, where they became highly successful. Some fast-swimming, shell-crushing marine predators, like the ichthyosaur (FIG. 3-5) preyed on the ammonites. It first punctured the shell from the blind side of the ammonite, causing it to fill with water and sink to the bottom. Then the attack could be made head on. These highly aggressive

Fig. 3-5. The ichthyosaur was one of the most spectacular marine reptiles of the Cretaceous seas.

predators might have been responsible for the extinction of most ammonite species prior to the end of the Cretaceous.

Lizards and turtles also went to sea in the Cretaceous. However, out of all the marine reptiles, only the smallest turtles made it past the extinction at the end of the period. The amphibians continued to decline throughout the Mesozoic and all large, flat-headed species became extinct early in the Triassic period. The group, thereafter, was represented by the more familiar toads, frogs, and salamanders.

When the Triassic closed, about 210 million years ago, large families of animals died in record numbers. The extinction which occurred over less than a million years, was responsible for killing nearly 50 percent of the reptile families. This occurrance forever changed the character of life on Earth and paved the way for the the rise of the dinosaurs.

Birds first appeared in the Jurassic period around 150 million years ago. *Archaeopteryx* (FIG. 3-6), which was about the size of a pigeon, is the earliest known fossil bird and appears to have been a species in transition between reptiles and birds. It was originally thought to be a small dinosaur until fossils showing impressions of feathers were found. This finding sparked a controversy over whether the feather impressions were real or etched into the fossil by someone. However, an Archaeopteryx fossil from a unique limestone formation in Bavaria produced a well-preserved specimen that clearly showed impressions of feathers. Although Archaeopteryx had all the accoutrements necessary for flight, it probably was a poor flyer and flew for only short distances.

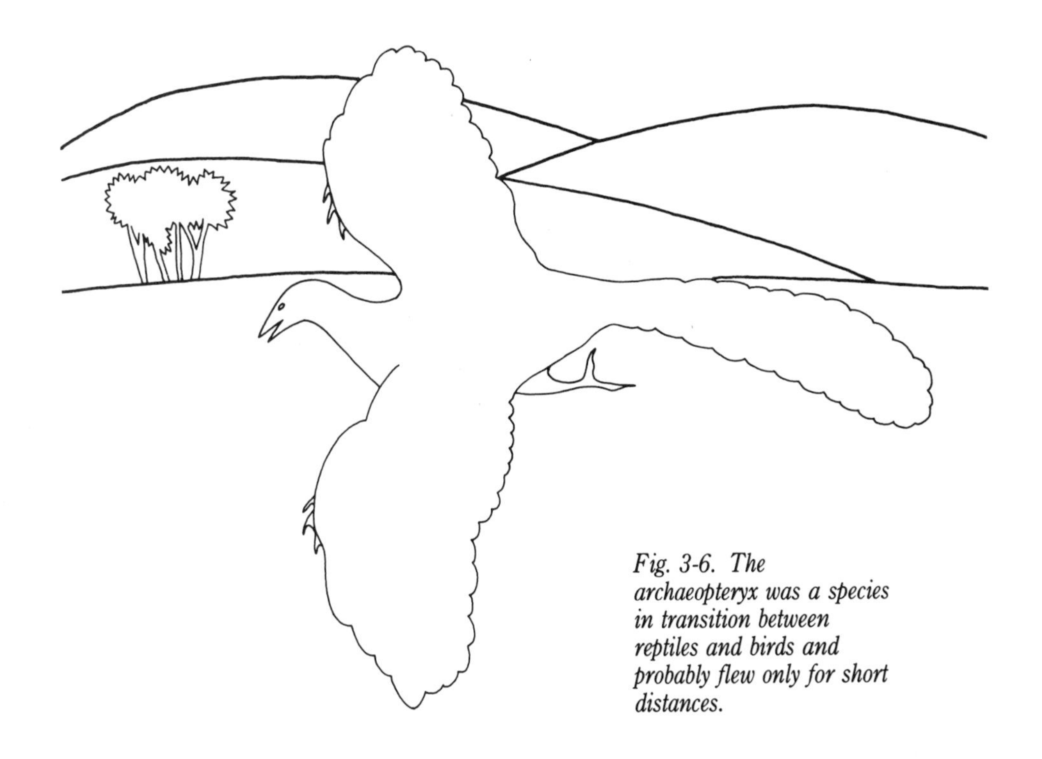

Fig. 3-6. The archaeopteryx was a species in transition between reptiles and birds and probably flew only for short distances.

After mastering the skill of flight, birds quickly radiated into all environments. Their better adaptability also allowed them to successfully compete with the flying reptiles, called *pterosaurs* (FIG. 3-7). This adaptability probably contributed to the extinction of the pterosaurs after these dinosaurs ruled the skies for over a hundred million years.

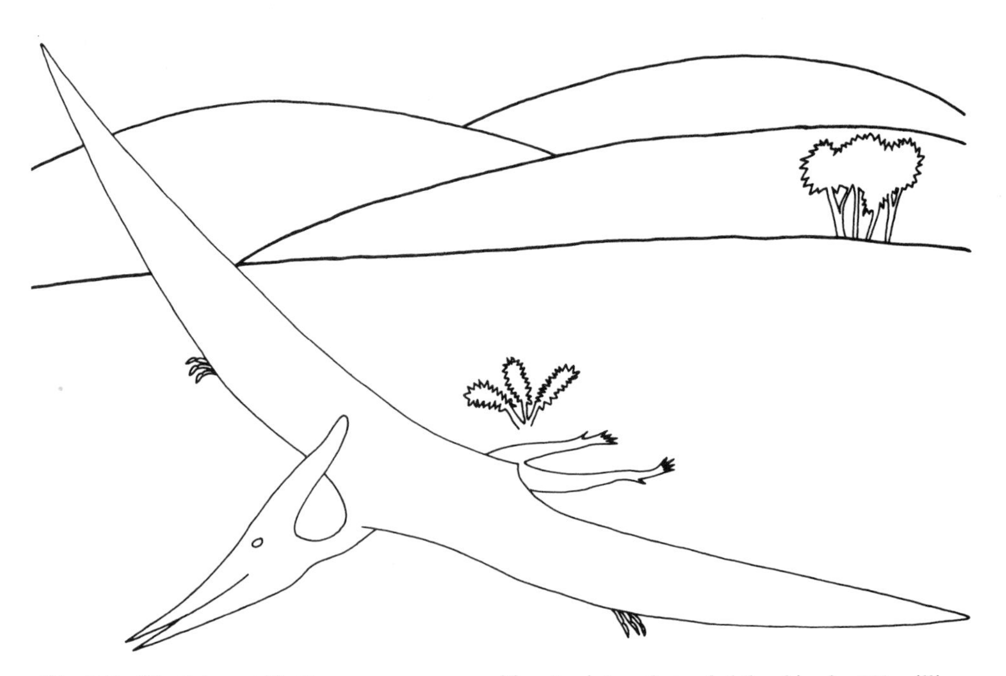

Fig. 3-7. The flying reptile pterosaurs were magnificant aviators that ruled the skies for 140 million years.

The pelycosaurs, which branched off from the reptiles about 300 million years ago, thrived for about 50 million years and then went extinct at the end of the Permian. After this extinction came their descendents, the mammal-like reptiles, called *therapsids.* After thriving for 40 million years, the therapsids were largely replaced by the dinosaurs. Thereafter, the ancestors of the mammals were probably relegated to being a shrewlike, nocturnal hunter of insects, until the dinosaurs themselves became extinct.

DINOSAUR EXTINCTIONS

The dinosaurs were the most successful land animals that have ever lived. They inhabited every corner of the globe and ruled the entire world for 140 million years. The success of the dinosaurs is exemplified by their extensive range, in which they occupied a wide variety of habitats and dominated all other forms of land-dwelling animals. Indeed, if the dinosaurs had not become extinct, mammals might not have achieved dominance over the Earth and humans might not have come into existence. The dinosaurs probably would have suppressed any further advancement of the mammals, forcing them to remain small, nocturnal creatures that tried to stay out of the way.

The oldest dinosaurs originated on the southern continent of Gondwana when the last of the glaciers from the great Permian ice age were disappearing. They ventured to all major continents, and their distribution throughout the world is evidence for continental drift. At the time the dinosaurs came into existence, all continents were assembled into a single huge landmass, called *Pangaea* (FIG. 3-8). During the Jurassic, which began about 180 million years ago, Pangaea began to rift apart and the present-day continents split off. With the exception of a few temporary land bridges, the oceans that filled the rifts between the newly formed continents provided an effective barrier to any further dinosaur migration.

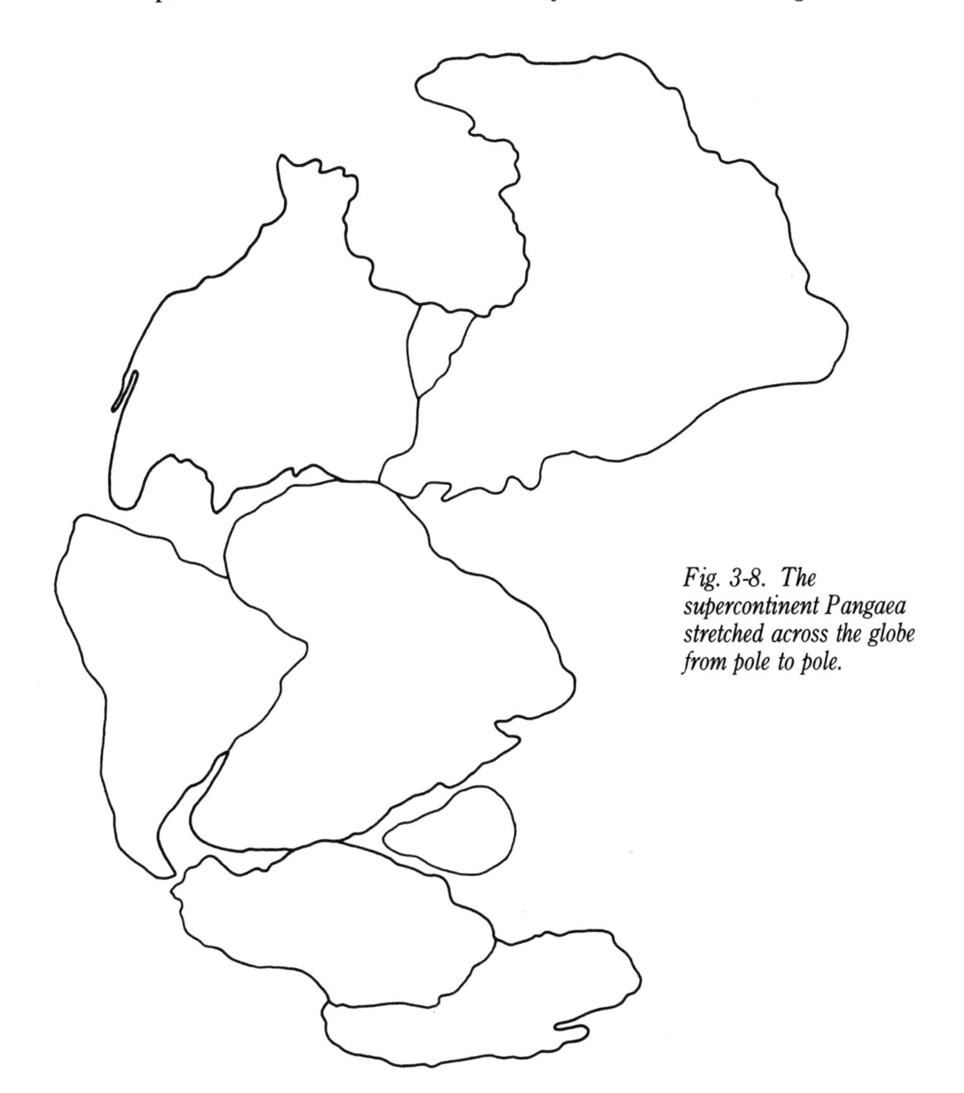

Fig. 3-8. The supercontinent Pangaea stretched across the globe from pole to pole.

The breakup of Pangaea might also have contributed to the extinction of the dinosaurs. The shifting continents changed global climate patterns and brought unstable weather conditions to many parts of the world. Massive lava flows, from what was perhaps the most volcanically active period since the early days of the Earth, might have dealt a major blow to the climatic and ecological stability of the planet.

Dinosaurs are often portrayed as unintelligent, slow-moving brutes. Yet, the fossil record seems to indicate that many species were swift as well as intelligent. The giant herbivores (FIG. 3-9) appear to have traveled in vast herds with the largest adults in the lead and the juveniles in the middle for protection. The females of some dinosaur species even might have given live birth. Many nurtured and fiercely protected their offspring until they could fend for themselves. This parental care of the young indicates strong social bonds and might explain why the dinosaurs were so successful for so long.

Fig. 3-9. Triceratops was a lumbering tank of a dinosaur that traveled in massive herds and were among the last to become extinct.

Paleontologists have classified about 500 species of dinosaurs, and many more are apt to be discovered. The generally warm climate of the Cretaceous allowed lush vegetation to grow throughout the world. Moreover, the abundance of food might have contributed to the giantism of some plant-eating dinosaurs. The largest of which, the 80-ton ultrasaurus, could look down on the roof of a five-story building. The angiosperms might also have contributed to the demise of the dinosaurs. Forests of broad-leaf trees and shrubs that were a favorite food of the dinosaurs are believed to have disappeared just prior to the end of the Cretaceous.

The extinction of 90 percent of the plankton at the end of the Cretaceous might have been caused by a major climate change, perhaps brought on by a massive meteorite shower. The impacts could have resulted in the widespread extinction of microscopic marine plants, called *calcareous nannoplankton.* These plants released a sulfur compound into the atmosphere, which aided cloud formation. The clouds in turn reflect sunlight and prevent solar radiation from reaching the surface. With the death of the calcareous nannoplankton, cloud cover would have been reduced. This lack of cloud cover could have triggered an extreme global heat wave, capable of killing off the dinosaurs and most marine species.

A collision with a large asteroid or a massive bombardment of meteorites also could have stripped away the ozone layer in the upper atmosphere, bathing the Earth in the Sun's deadly ultraviolet rays. These rays would have killed land plants and animals, as well as primary producers in the surface waters of the ocean. Because mammals were mostly nocturnal and remained in their underground borrows during the day, only coming out at night to feed, they would have been spared from the onslaught of ultraviolet radiation. Thus protected, they were able to replace the "naked" dinosaurs.

This scenario has important implications for us today. For if we do not halt the destruction of the ozone layer with our pollutants, today's "asteroid collision" might be man-made.

RECENT EXTINCTIONS

Most of the Paleocene epoch, which began about 65 million years ago, was characterized by an evolutionary lag, as though the world had not yet fully recovered from the great extinction at the end of the Cretaceous. Nevertheless, toward the end of the Paleocene, the mammals began to diversify rapidly, and some large, peculiar-looking animals became evolutionary dead-ends.

The mammals were cut down again during a sharp extinction event at the end of the Eocene epoch, about 37 million years ago, when the Earth took a plunge into a colder climate. At this time, many of the archaic mammals abruptly disappeared. Of the dozen or so orders of mammals that existed in the early Cenozoic era, only half were found in the proceeding Cretaceous and only half are alive today. Only in the Eocene, which began about 54 million years ago, did truly modern mammals began to immerge (FIG. 3-10).

Marine species that survived the extinction event at the end of the Cretaceous looked much the same in the Cenozoic as they did in the Mesozoic. The ocean has a moderating effect on evolutionary processes because it has a longer memory of environmental conditions than the land—it requires much more time to heat or cool. The ocean also provides a better protective shield than the atmosphere against cosmic rays and ultraviolet radiation. Therefore, the rate of genetic mutations is slower in the seas than it is on land.

Major marine groups that disappeared at the end of the Cretaceous include the ammonites, the rudists (huge coral-shaped clams), and other types of clams and oysters. All of the cephalopods were absent in the Cenozoic seas, except for the nautilus and the shell-less species, including cuttlefish, octopi, and squids. The squids competed directly with fish, which themselves were little affected by the

Fig. 3-10. Fossil mammals on display at the Museum of Geology, South Dakota School of Mines at Rapid City.

extinction. The gastropods, which include snails and slugs, increased in number and variety throughout the Cenozoic, and presently they are second only to insects in diversity.

During the Pleistocene glaciation, which began about 2.4 million years ago, adaptations to the cold climate allowed certain species of large mammals to thrive in the glacier-free north lands. Giant mammals, including mammoths, mastodons, saber-toothed cats, and ground sloths that reared 20 feet tall, inhabited many parts of the Northern Hemisphere.

When the glaciers began to retreat around 16,000 years ago, major readjustment in the global environment occurred as the cool, equable climate of the ice age gave way to the warmer, more seasonal climate of our present interglacial. The rapid change from a glacial to an interglacial climate shrank the forests and expanded the grasslands. This change might have disrupted the food supplies of several large mammals. Deprived of their nutritional resources, they simply disappeared.

PREHISTORY EXTINCTIONS

The earliest humans, called *Australopithecus* (FIG. 3-11), were markedly different from us. They first appeared about four million years ago in Tanzania

(COURTESY OF NATIONAL MUSEUMS OF CANADA)

Fig. 3-11. The human ancestry probably began with Australopithecus.

and Ethiopia, and perhaps even earlier in northern Kenya. One remarkable discovery of an almost half-complete skeleton of a female Australopithecus named "Lucy" was found at Hadar, Ethiopia and was between three and four million years old. These hominids were primitive in most of their features. They were generally three to four feet tall and weighed less than 100 pounds. Moreover, the males were about twice as large as females, a typical characteristic of human species known as *sexual dimorphism.*

Around 2.4 million years ago, the climate cooled and initiated a series of ice ages. It also prompted a shift toward the more open, savannalike habitats in Africa. These conditions resulted in the appearance of many new animal species and spurred the development of early humans. At this time, probably several species of African hominids lived in the same area, and it is possible that they preyed upon each other. The biggest hominid, called *Gigantropithecus,* was larger than a gorilla. It was over six feet tall and weighed nearly four hundred pounds. It was also an evolutionary dead-end and became extinct about 500 thousand years ago.

A little over two million years ago, a larger-brained hominid arose, called *Homo habilis.* This first line of the human ancestry was much like Australopithecus in its face and teeth, but it had a significantly larger brain, which averaged about 700 cubic centimeters, about half the size of modern human brains. The limb bones were unlike those of Australopithecus and resembled those of later human species. The two hominid lines coexisted in Africa until almost one million years ago, at which time Australopithecus became extinct.

It is generally thought that *Homo erectus* was a direct descendent of Homo habilis. It was the first widely distributed hominid, appearing in Africa about 1.5 million years ago. 500 thousand years later, the species was present in Southern and Eastern Asia, where it lived until a little over 200 thousand years ago. It has also been suggested that Homo erectus originated in Asia and migrated to Africa, implying that it evolved independently of Homo habilis.

Beginning about 130,000 years ago, during the last interglacial period, the Neanderthals ranged over Western Europe and Central Asia and thrived in those areas until about 35 thousand years ago. Their great success is illustrated by the fact that they were able to endure the rigors of the cold climate during the last ice age. The activity of the Neanderthal's massive muscles probably supplied their bulky bodies with the heat necessary to survive these bitter cold conditions. Other physical attributes might also have been adaptations to the cold subarctic conditions, similar to those exhibited by modern Eskimos and Lapps.

The sudden departure of the Neanderthals after some 100,000 years of prosperity might have resulted from stiff competition with a more advanced human species, known as *Cro-Magnons.* These modern humans originated in Africa some 90,000 years ago, but for some reason, they did not penetrate further into the Old World for another 50,000 years. By this time, Neanderthals were scattered throughout Eurasia (FIG. 3-12). The demise of the Neanderthals coincides with the rise of the Cro-Magnon, which suggests that the latter might have killed their stocky cousins in a sort of Stone Age holocaust.

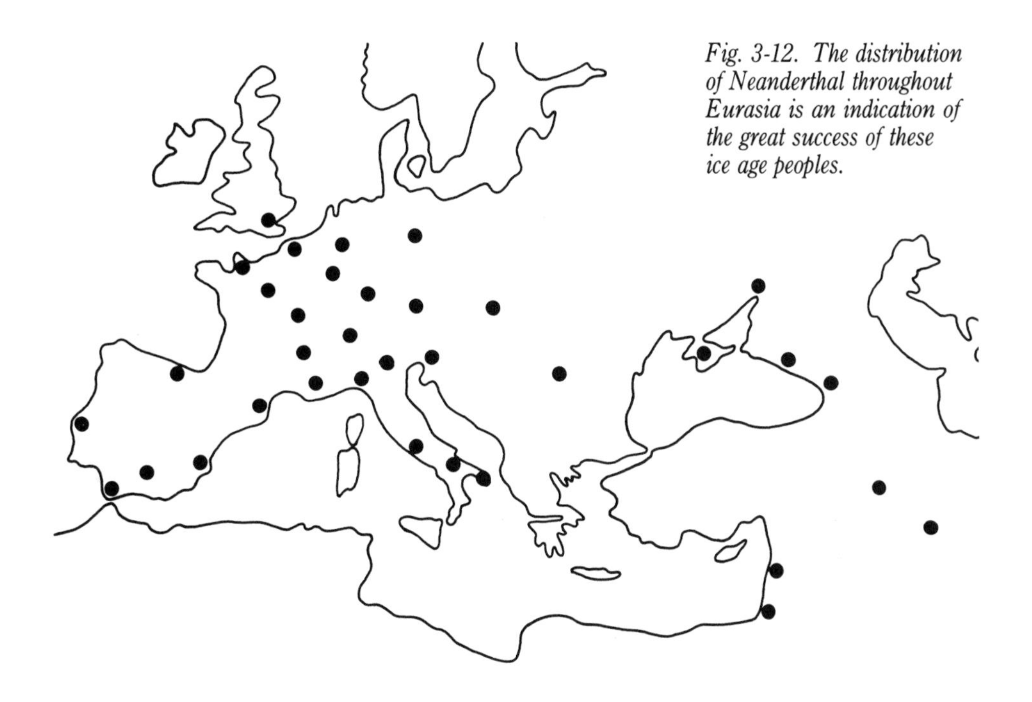

Fig. 3-12. The distribution of Neanderthal throughout Eurasia is an indication of the great success of these ice age peoples.

4

Evolution of Species

THERE are two types of classical evolutionary evidence. One is information on living organisms as they exist today and the other is fragmentary information on earlier life as read in the fossil record. The history of the Earth as told by its fossil record is not complete, however, because of the remaking of the surface, which erases entire chapters of historic geology. Yet the study of fossils, along with the radiometric dating of the rocks that contain them, has constructed a reasonably good chronology of Earth history. Thus, for the most part, the fossil record provides clear evidence for the evolution of life on Earth.

Geologists measure geologic time by tracing fossils through rock strata and noticing the greater evolutionary changes with rocks lower down, as compared to those higher up in the geologic column. Fossil-bearing strata can also be followed horizontally over great distances because a particular fossil bed can be identified in another locality, with respect to the beds above and below it. These *marker beds* are important for use in identifying and dating geologic formations.

When fossils are arranged chronologically, they do not present a random or haphazard picture. Rather, they show progressive changes from simple to more complex life-forms and reveal the evolutionary advancement of species through time. Paleontologists are thus able to recognize geologic time periods based on groups of organisms that were especially plentiful and characteristic during a particular time period. Within each period are many subdivisions, determined by the occurrence of certain species. This same succession is found on every major continent and is never out of order.

THE TREE OF LIFE

Extinct organisms are classified by the same system used to classify *extant*—living organisms. The first classification scheme was developed by the eighteenth century Swedish botanist, Carl von Linne, better known as Carolus Linnaeus. He named his organisms using Latin words because Latin was the universal language of science in his day. His naming was based on the number of characteristics that organisms had in common; for example, birds, bats, and pterosaurs all could fly.

During his studies, Linnaeus began to realize that some organisms were more similar than others because they were more closely related. Later, as evolution became recognized as the process by which organisms develop into new species, classification schemes were developed to describe these evolutionary patterns. These classifications demonstrated how groups of organisms were related, both spatially and chronologically.

Each organism is assigned an italicized, two-part species name. The first word, which is capitalized, is the generic name and is shared with other closely related species. The second word, which is not capitalized, is the species name and is unique to a particular genus. For example, our species name is *Homo sapiens.*

Sometimes the discoverer's name and the date of discovery will follow the species name. The scientific name of a species is written either in Greek or Latin so there will be enough names for the well over two million known species and the hundred or so new species that are discovered every day. All species must have a unique name—it cannot be used again for another species, nor can more than one name be used for the same species.

The classification of taxa establishes a hierarchy of organisms (TABLE 4-1). Each step up the evolutionary scale becomes more inclusive, encompassing a larger number of species. A *kingdom* is the largest taxonomic classification and comprises all the species of either animals, plants, or microorganisms. Our kingdom is Animalae, which comprises all the animals. Our *phylum* is Chordata, which comprises all the vertebrates. Our *class* is Mammalia, which comprises all the mammals. Our *order* is Primate, which also comprises monkeys and apes. And of *family* is Hominidea, which comprises all humans. Genus and species are the

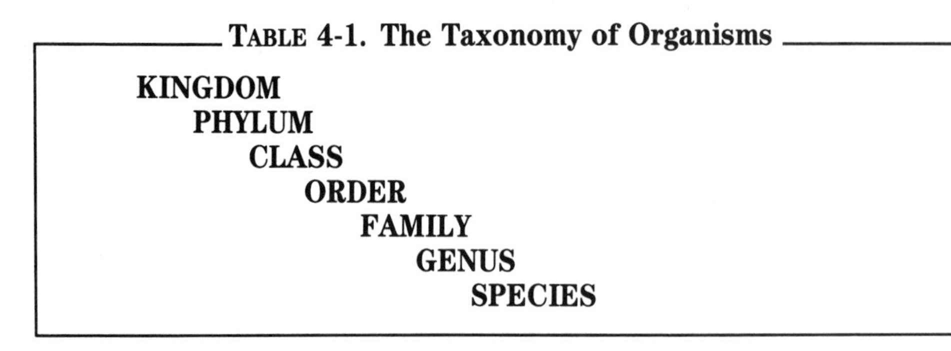
TABLE 4-1. The Taxonomy of Organisms

KINGDOM
PHYLUM
CLASS
ORDER
FAMILY
GENUS
SPECIES

lowest rungs of the classification ladder. Our genus, Homo, also encompasses all of our primitive ancestors, beginning with Homo habilis, which lived about two million years ago.

Just 10 phyla are needed to classify the vast majority of animal life on Earth both past and present (TABLE 4-2). The first phylum, Protozoa, begins with the simplest life-forms, and each succeeding phylum becomes more complex. The phylum Chordata, where we belong, contains the most complex life-forms. The phyla have been organized in this manner to recognize the evolutionary advancement of species.

The first organisms to develop were probably bacteria and ancestral algae. On the next stage in the evolutionary scale are the *protozoans*, including the amoeba, foraminifera, and radiolarian. The next step up contains the *sponges*, which belong to the phylum, Porifera. The *coelenterates* include jellyfish, sea anemones, hydra, and coral. The calcareous coral built impressive formations of limestone, which trapped and fossilized other organisms. More recent corals are responsible for the construction of barrier reefs and atolls.

The *bryzoans*, or moss animals, attached to the seafloor and filter fed on microscopic organisms. They are important marker fossils for correlating rock formations over long distances. The *brachiopods* are among the most common fossils with over 30,000 separate species cataloged in the fossil record. Perhaps these were the shells that baffled the ancient Greeks when they discovered them high in the mountains, far from the seashore.

The *molluscs*, which include snails, clams, and squids, probably left the most impressive fossil record of all marine species. The extinct ammonites possessed giant spiral shells, some several feet in diameter. The *annelids*, which include worms and leaches, left behind a profusion of tracks and borings (FIG. 4-1). The largest phylum of living organisms are the *arthropods*, which includes insects, spiders, shrimp, lobsters, and barnacles. Perhaps the first and best-known arthropods were the extinct trilobites.

The *echinoderms*, which means spiny skin, are possibly the strangest animals ever preserved in the fossil record. They are unique because they possess radial symmetry and a water vascular system used for feeding and locomotion. They also have no head. Included in this phylum are sea lilies, sea cucumbers, starfish, brittle stars, and sea urchins.

The higher animals are the *vertebrates*, which include fish, amphibians, reptiles, birds, mammals, and humans. The descendents of the lobed-fin fish and lungfish were the first advanced animals to populate the land 370 million years ago. Their legacy is well-documented in the fossil record, and at no other time in geologic history were so many varied and unusual creatures inhabiting the surface of the Earth.

EVIDENCE FOR EVOLUTION

The chief process by which evolution progresses is through natural selection, an essentially ecological process based on the relationships between organisms and their environments. Certain inherited traits allow species to become particu-

Table 4-2. Classification of Species

Group	Characteristics	Geologic Age
Protozoans	Single celled animals, Forams and radiolarians.	Precambrian to recent
Porifera	The sponges, about 3000 living species.	Proterozoic to recent
Coelenterates	Tissues composed of three layers of cells. About 10,000 living species. Jellyfish, hydra, coral.	Cambrian to recent
Bryozoans	Moss animals. About 3000 living species.	Ordovician to recent
Brachiopods	Two asymmetrical shells. About 120 living species.	Cambrian to recent
Molluscs	Straight, curled, or two symmetrical shells. About 70,000 living Species. Snails, clams, squids, ammonites.	Cambrian to recent
Annelids	Segmented body with well-developed internal organs. About 7000 living species. Worms and leaches.	Cambrian to recent
Arthropods	Largest phylum of living species with over one million known species. Insects, spiders, shrimp lobsters, crabs, trilobites.	Cambrian to recent
Echinoderms	Bottom dwellers with radial symmetry. About 5000 living species. Starfish, sea cucumbers, sand dollars, crinoids.	Cambrian to recent
Vertebrates	Spinal column and internal skeleton. About 70,000 living species. Fish, amphibians, reptiles, birds, mammals.	Ordovician to recent

(PHOTO BY D.L. SCHMIDT, COURTESY OF USGS)

Fig. 4-1. Often the only evidence of an animal's existence is revealed by its comings and goings, such as shown in these worm borings in sandstone from the Pensacola Mountains, Antarctica.

larly well-suited to survive and reproduce in their prevailing environment. During environmental stress, species that acquire favorable traits through mutations adapt more easily and are more likely to survive and pass these survival characteristics to their offspring.

Since there are many different environments, a wide variety of species exist. Therefore, evolutionary trends have varied throughout geologic time, in response to major environmental changes, as natural selection adapts organisms to the new conditions forced on them by a number of environmental factors. However, natural selection is not deterministic. Variations are purely accidental and are selected in accordance with the demands of the environment. Most of the time, species resist changes, even when these changes make them better fitted to environmental needs.

One of the ways to understand evolution on a small scale is by observing species in transition. During the last century in central England, prior to the advent of coal-burning factories, many species of moths were camouflaged against light-colored tree trunks and were protected from attacks by birds. Darker moths that were produced by occasional mutations were more visible to the birds and therefore were at a disadvantage. But as coal soot blackened the tree trunks, the darker moths were no longer conspicuous and the lighter moths became easy prey. In a mere 30 to 50 years, a new race of dark moths evolved that had a distinct advantage over their lighter-colored relatives.

Similarly, unwanted pests and weeds develop resistences to insecticides and herbicides almost as rapidly as new chemicals are developed for their eradication. Some vegetation that grows around industrial plants have evolved forms that are resistant to the toxic chemicals and heavy metals that end up in the soil. As we change the chemistry of the environment with our pollution, we are inadvertently causing new and potentially dangerous species to evolve.

The long neck of the giraffe is often cited as a classic example of evolutionary adaptation (FIG. 4-2). In this case, a browsing animal, whose diet was tree leaves, responded to a food supply that grew increasingly higher. The trunk of the elephant is another adaptation for browsing, but in this case the snout was elongated to reach tall branches and to dig for water during dry spells. There are two living species of elephants, one of which is closely related to the extinct mammoth. Mastodons were also related, and diverged from the elephant family about 30 thousand years ago.

Another example of evolution at work is the horse, which shows a clear picture of evolutionary change through time. The earliest horses originated in western North America during the Eocene epoch and were about the size of a small dog. As time progressed, horses became progressively larger. Their faces and teeth grew longer as the animal switched from browsing to grazing, and a single toe on each foot evolved into a hoof. Many types of hoofed animals, called *ungulates* developed in response to increasing grasslands that covered many parts of the world.

GENETIC MUTATION

The means by which organisms are able to adapt to their changing environment is through genetic mutation. Altered genes in the chromosomes inact certain changes that allow a species and its offspring to live more successfully in their surroundings. Mutations occur as a direct result of the chemistry of the environ-

Fig. 4-2. The evolution of the giraffe is clear evidence of adaptation from a grazing to a browsing life-style.

ment and the bombardment of genes by cosmic and background radiation. This theory also signifies that the environment has direct control over the rate of mutation. The mutation can be either beneficial or harmful to the organism and thus determines who survives and who perishes.

For years, scientists have held that the only organisms able to survive environmental upheavals, such as abrupt temperature shifts or food shortages, are those that have mutated before encountering the stress. In other words, those species that changed into a more adaptive form prior to the upheaval were better able to withstand abrupt environmental stress. However, recent experiments on bacteria might force scientists to revise this view. Bacteria can somehow adopt genetic traits in response to a particular environment. They then pass on these acquired traits to their offspring, which then have a better chance of survival. Mutations in bacteria arise spontaneously and randomly, giving them the ability to mutate in a more purposeful manner in order to adapt to a particular environment.

When a mass extinction occurs, however, it selects for survival those individuals that have already been altered, rather than being the direct cause of the mutations. This is why certain species are able to survive one mass extinction after another. It is particularly true of marine species like the sharks (FIG. 4-3), which began in the Devonian, around 400 million years ago, and have survived every mass extinction since that time. Unfortunately, heavy hunting pressures by its only natural enemy, man, might finally cause the demise of the sharks after they had been so successful for such a long time.

Fig. 4-3. The sharks are fierce predators that evolved in the Devonian seas.

SURVIVAL OF THE FITTEST

During the globe-hopping journey of the HMS *Beagle*, from 1831 to 1836, the English naturalist Charles Darwin was employed as the ship's geologist. He described in great detail the rocks and fossils he encountered on his trip around the world (FIG. 4-4). It is interesting to note that Darwin was trained as a geologist and thought as a geologist, but today he tends to be viewed as a biologist. He made many significant contributions to the field of geology, which during his day was entering its golden age.

On his voyage, Darwin observed the relationships between animals on islands and on adjacent continents as well as between animals and fossils of their extinct relatives. This study led him to the conclusion that species were continuously evolving through time (FIG. 4-5). Therefore, to Darwin, evolution worked at a constant tempo as species adapted to a constantly changing environment. Actually, Darwin was not the first to make this observation. Where his theory differed, however, was in postulating that new parts evolved in many tiny stages, rather

Fig. 4-4. Darwin's journey around the world from England to South America, Australia, Africa, and back.

Fig. 4-5. The stages in the development of reptiles from the late Devonian Ichthyostega (top) to the late Paleozoic labyrinthodont (middle) to the middle Pennsylvania hylonomus (bottom).

than in discrete jumps. He therefore attributed gaps in the fossil record to erosion or nondeposition of fossil-bearing strata.

Darwin coined the phrase "survival of the fittest," which signifies that the members of a particular species that can best utilize their environment have the best chance of producing offspring that possess the survival characteristics of their parents. In other words, successful parents have a better chance of passing on their "good" genes to their offspring, which in turn are better able to survive in their respective environments. Natural selection, therefore, favors those best-suited to their respective environments at the expense of weaker species.

Contemporary geologists embraced Darwin's theory, for at last there was a clear understanding of the evolutionary changes in fossils of different ages. With this knowledge, they were able to piece together geologic history by placing events in their proper sequence through the study of the evolutionary changes in fossils. Only the fossils of extinct species can testify to the historical patterns of evolution traced through geologic time.

Evolution has not always been a gradual and constant process, as Darwin saw it, however. The fossil record seems to indicate that life evolved by fits and starts. Many long periods of little or no change are punctuated by short periods of rapid change, followed again by long periods of stasis. Rapid evolutionary changes in large segments of organisms sometimes appear in the fossil record as the result of a mass extinction when in fact no real extinction had occurred.

Rapid evolution might also result from rare large mutations. Thus, when evolution occurs, it will be by a sudden leap, with major changes occurring in many parts. In other words, natural selection does not favor piecemeal tinkering. Natural selection, therefore, cannot work on structures that are not fully functional during intermediary periods of development of new body parts. An example is the development of insect wings, which were probably first used to cool the animal. Later, as the benefits of flight were felt, they became more aerodynamic, giving flying insects a huge advantage.

Evolution might also be opportunistic, in that variations arise by chance and are selected in accordance with the demands of the environment. When the environment changes abruptly to one that is harsher, species that cannot adapt to these new conditions fast enough do not live at their optimum. Therefore, they do not pass on their "bad" genes to future generations.

One of the most frustrating aspects of the geologic time scale is the gaps in the fossil record, where the history of the Earth has been erased. This might have been caused by periods of erosion or the nondeposition of sedimentary strata that trapped and preserved species as fossils (FIG. 4-6). Gaps in the fossil record might also be the result of insufficient intermediary species, or so-called missing links, which might have existed only in small populations. Small populations are less likely to leave a fossil record than larger ones because the process of fossilization favors large populations.

Moreover, the intermediates might not have lived in the same locality as their ancestors and thus were unlikely to be preserved along with them. New species that start in small populations might evolve rapidly as they radiate into new

(PHOTO BY H.E. MALDE, COURTESY OF USGS)

Fig. 4-6. Fossil shells of mollusks near Glenns Ferry, Idaho.

environments. Then as populations increase, slower evolutionary changes occur as the species' chances of entering into the fossil record improve.

The origination and extinction of species throughout the fossil record is extremely important in building an accurate account of the evolution of species through time. However, some statistical traps in the fossil record might suggest differences in fossil samples where no actual differences exist. In any ecological community, a few species will occur in abundance, some will occur frequently, but most are rare and occur only infrequently. Moreover, the chance of an individual becoming fossilized after death and thus entering the fossil record is extremely small.

Another major problem lies in the method of sampling. For instance, no single fossil sample will contain all the rare organisms in an assemblage of species. If one particular sample was compared with another higher in the stratigraphic column, which represents a later time in geologic history, an overlapping but different set of rare species would be found. Species found in the lower sample, but not in the upper sample, might erroneously be inferred to have gone extinct. Conversely, species that appear in the upper sample, but not in the lower sample, might wrongly be thought to have originated there. Thus, the reading of the fossil record can be confusing and misleading.

CONTINENTAL DRIFT

The continents have been drifting over the face of the Earth almost since the crust first formed four billion years ago. This is manifested in four-billion-year-old granites discovered in Canada's Northwest Territories, which suggest that the formation of the crust was well underway by this time. Among the strongest forces that affect evolutionary changes is *continental drift.* The motions of the continents have a wide-ranging effect on the distribution, isolation, and evolution of species. The changes in continental configuration greatly affect global temperatures, ocean currents, productivity, and many other factors of fundamental importance to life.

The positioning of the continents, with respect to each other and to the equator, help determine climatic conditions. When most of the land existed near the equator, the climate was warm. However, when lands wander into the polar regions, the climate turns cold and spawns episodes of glaciation and mass extinction.

The movement of the continents changes the shapes of ocean basins. This change affects the flow of ocean currents, the width of continental margins, and the abundance of marine habitats. When a supercontinent breaks apart, more continental margins are created, the land lowers, and the sea level rises, providing a larger habitat area for marine organisms.

Greater volcanic activity occurs during times of highly active continental movements. This is especially true at midocean spreading centers, where crustal plates are pushed apart by upwelling magma from the upper mantle, and at subduction zones, where crustal plates are forced into the Earth's interior and remelted to provide the raw materials for a new crust (FIG. 4-7). The amount of volcanism could affect the composition of the atmosphere, the rate of mountain building, and the climate—all of which inevitably affect life.

The drifting of the continents during the Mesozoic era isolated many groups of mammals and they evolved along independent lines. Australia is an excellent example of this and is inhabited by strange egg-laying mammals, called *monotremes,* including the spiny ant eater and the platypus, both of which should rightfully be classified as surviving "mammal-like reptiles." When the platypus was first discovered, it was thought to be the missing link between mammals and their forbearers.

Marsupials are believed to have originated in North America around 100 million years ago. They then migrated to South America, crossed Antarctica when the two continents were still attached to each other, and landed in Australia before it broke away from Antarctica around 40 million years ago (FIG. 4-8). The Australian group is composed of kangaroos, wambats, bandicoots, etc., and opossums and their relatives occupy other parts of the world.

Camels, which originated about 25 million years ago, migrated out of North America to other parts of the world by crossing over connecting land bridges

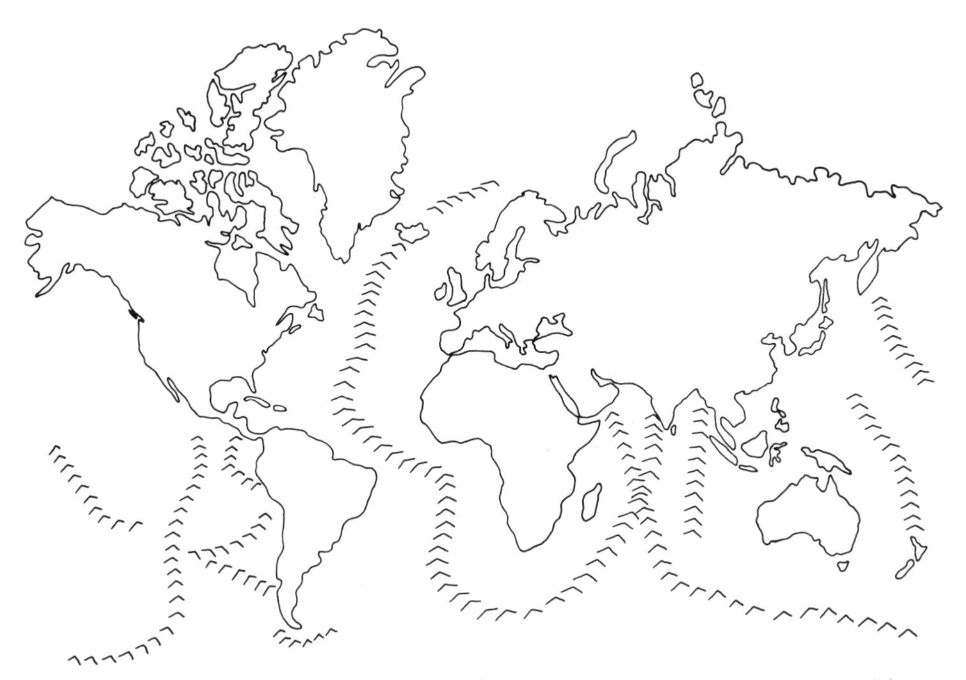

Fig. 4-7a. Midocean ridges, where crustal plates are spreading apart, are centers of intense volcanic activity.

Fig. 4-7b. Subduction zones, where crustal plates are forced into the Earth's interior, are other regions of intense volcanic activity.

Fig. 4-8. The migration route of marsupials to Australia via Antarctica.

(FIG. 4-9). Madagascar, which broke away from Africa about 125 million years ago, has none of the large mammals that occupy the mainland—except for the hippopotamus, which mysteriously landed on the island after it had drifted far from the mainland.

THE GAIA HYPOTHESIS

Generally it is believed that environmental change drives evolution. However, in 1979, the British chemist James Lovelock turned this argument around by

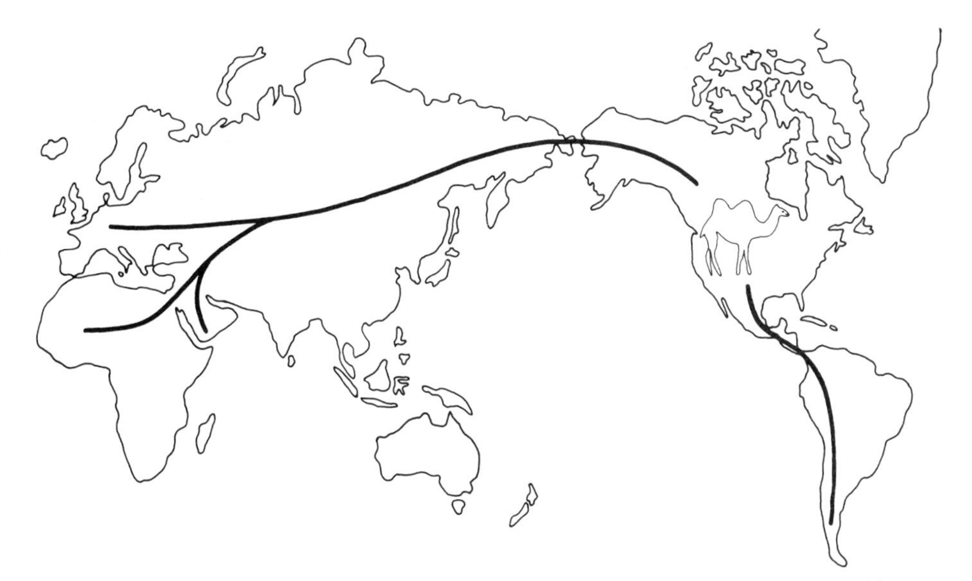

Fig. 4-9. The dispersion of camels from North America to other parts of the world.

proposing the so-called *Gaia hypothesis*, named for the Greek goddess of the Earth. It postulates that life has a certain degree of control over its environment, and that the biosphere maintains the optimum living conditions for all organisms by regulating the climate. For example, a certain species of plankton releases a sulfur compound into the atmosphere, which aids in the formation of clouds. If the Earth warms up, it invigorates the plankton, which release more cloud-forming sulfur compounds. These compounds produce more clouds, which cool the planet and thus stabilize the temperature.

The Gaia hypothesis suggests that from the very beginning, life followed a well-ordered pattern of growth, advancing from simple to complex organisms, independent of chance and natural selection. It seems that life has kept pace with all the major changes in the Earth over time and might have made some alterations of its own. One major change was the conversion of the atmosphere from nearly 25 percent carbon dioxide to nearly 25 percent oxygen as a result of photosynthesis. Plants use the carbon dioxide to manufacture organic compounds, and produce oxygen as a byproduct.

As life progressed, slow but steady changes greatly affected the final outcome of the planet. Like the Earth, the other planets and their satellites have a core, a mantle, a crust, and even an atmosphere or an ocean of sorts, but none of them have a biosphere. Moreover, a biosphere requires more than just living entities. Life must also be integrated with the lithosphere, hydrosphere, and atmosphere to constitute a fully developed biosphere.

Since the beginning, life has responded to a variety of chemical, climatological, and geographical changes in the Earth, that have forced species to either adapt or perish. Many dead-end streets along the branches of the evolutionary tree

Fig. 4-10. Geological time spiral depicting the evolutionary stages of the Earth. (COURTESY OF USGS)

are found in the fossil record, which itself represents only a fraction of all the species that have ever existed. Just about every conceivable form and function has been tried; some have been more successful than others. It is through this trial-and-error method of specialization that some species have prospered and less adaptive species have become extinct.

Life constitutes a geological force that no other planet has. What appear to be the earliest fossilized remains of stromatolite microorganism structures go as far back as 3.5 billion years. The 3.8-billion-year-old carbonaceous sediments of the Isua Formation in southwest Greenland show a depletion of carbon-13 with respect to carbon-12, which is thought to be a common manifestation of biological activity. Therefore, life processes appear to have been operating for at least 80 percent of Earth's history (FIG. 4-10).

Over such a lengthy period, life caused many changes in the Earth. The first major alteration was the production of massive iron deposits on continental margins, when photosynthetic organisms evolved and oxidized the iron in the ocean. Biological processes are also responsible for massive concentrations of other minerals, including silicon, carbon, manganese, copper, and sulfur.

A second major change in the Earth was the production of oxygen when photosynthesis evolved as an energy source. The oxygen produced a secondary benefit by forming the ozone layer in the upper stratosphere. This layer made it possible for plants and animals to leave the ocean and populate the land. It also helps solve one of the great riddles about the Earth; namely, why did life, which had been around for most of Earth's history, take so long to finally conquer the land?

5

Effects of Extinctions

SINCE the dawn of life, species have become extinct, paving the way for the development of new species that might better utilize the environment. For this reason, extinctions play an important role in the evolution of life. When a mass extinction occurs, new species develop to replace those that died out. If species did not become extinct to make room for more advanced organisms, life on Earth would not have progressed to where it is today. The only organisms would be simple microscopic creatures in the sea, the same as when life started.

Geologists are beginning to accept global catastrophes, such as mass extinctions, as normal occurrences in geologic history. Thus extinctions play an important role in the *uniformitarian process,* which postulates that the slow changes observed on Earth today had their counterparts in the geologic past. Certain periods of mass extinctions, however, appear to be the result of catastrophic events, rather than the usual modes by which extinctions occur (such as subtle changes in climate or sea level or an increase in predation). Therefore, mass extinctions appear to have played a prominent role in the pattern of life throughout most of the Earth's history.

MAJOR EXTINCTIONS

Throughout geologic history, species have regularly come and gone (TABLE 5-1), and many have disappeared in large numbers at particular junctures in time. The number of different species living on Earth today represents less than one

TABLE 5-1. Radiation and Extinction for Major Organisms

ORGANISM	RADIATION	EXTINCTION
Marine invertebrates	Lower Paleozoic	Permian
Foraminiferans	Silurian	Permian & Triassic
Graptolites	Ordovician	Silurian & Devonian
Brachiopods	Ordovician	Devonian & Carboniferous
Nautiloids	Ordovician	Mississippian
Ammonoids	Devonian	Upper Cretaceous
Trilobites	Cambrian	Carboniferous & Permian
Crinoids	Ordovician	Upper Permian
Fishes	Devonian	Pennsylvanian
Land plants	Devonian	Permian
Insects	Upper Paleozoic	
Amphibians	Pennsylvania	Permian-Triassic
Reptiles	Permian	Upper Cretaceous
Mammals	Paleocene	Pleistocene

percent of all those that have gone before. Billions of species of plants and animals are believed to have occupied the Earth at one time or another. Most lived during the Phanerozoic eon, the last 570 million years. During the Cenozoic era, species diversity reached one of its highest peaks so that today our world is filled with a huge variety of plants and animals (FIG. 5-1).

The number of species that are believed to exist today range from five million to as many as 30 million. Most of them live completely unnoticed by man. Many species play a critical role in food chains and make important nutrients available to higher organisms. It is these simple creatures, including bacteria, fungi, and plankton, that compose 80 percent of the Earth's biomass—the total weight of all living matter. Moreover, marine phytoplankton produces most of the breathable oxygen available on the planet.

The first major mass extinction occurred in the late Precambrian, about 670 million years ago. At that time, animal life was still sparse, and the extinction decimated the ocean's population of single-celled phytoplankton, which were the first organisms to develop cells with nuclei. The mass disappearance of this species coincided with a period when glaciers covered many parts of the world. When the ice disappeared near the end of the Precambrian, a mass diversity of new species appeared, the likes of which have never been seen before or since.

A second mass extinction occurred at the end of the Ordovician period (about 440 million years ago), which eliminated 100 families of marine animals. Another major extinction occurred near the end of the Devonian period, about 370 million years ago, in which many tropical marine groups simultaneously disappeared. The largest mass extinction occurred at the end of the Paleozoic era, about 240 million years ago, when over 95 percent of all marine species vanished (FIG. 5-2). This event was followed by another major mass extinction about 210 million years ago, when nearly 50 percent of the reptilian species disappeared. The most well-known mass extinction was that of the dinosaurs, along with 70 percent of all known species, at the end of the Cretaceous period.

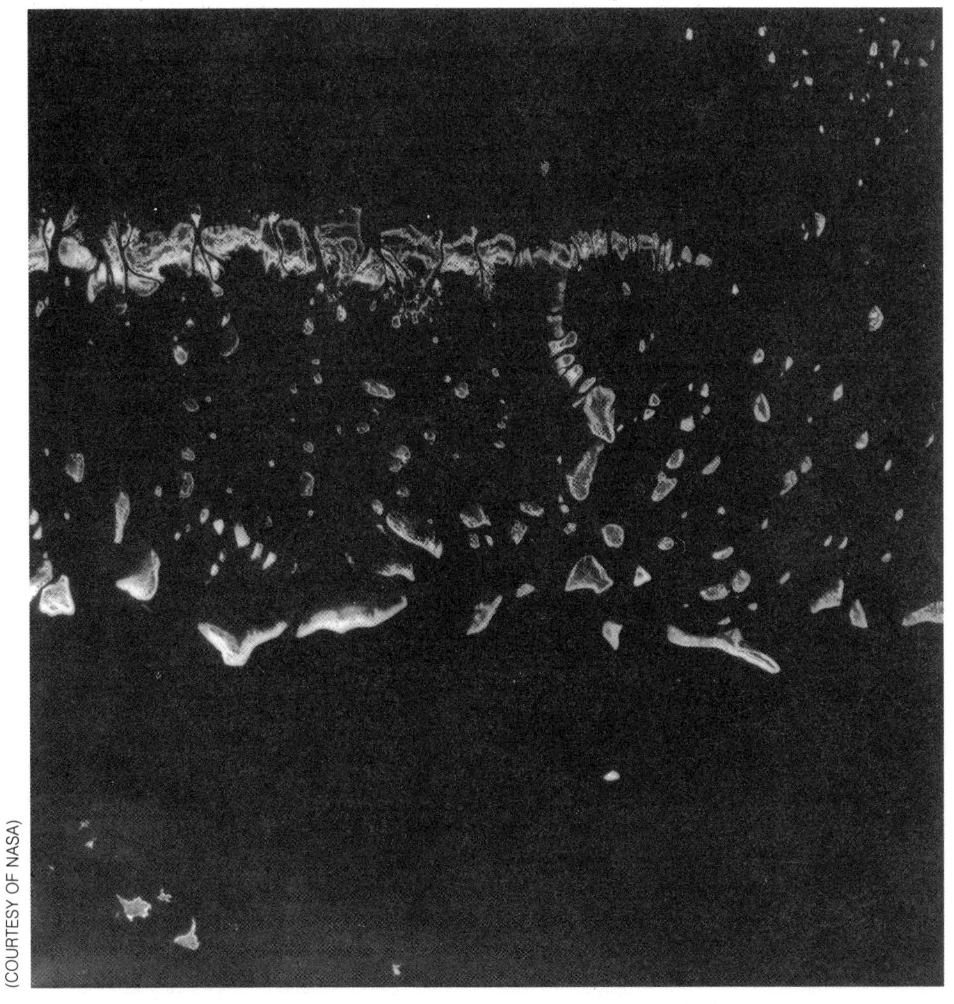

(COURTESY OF NASA)

Fig. 5-1. The Great Barrier Reef is the world's most extensive reef system and home to a large variety of marine species.

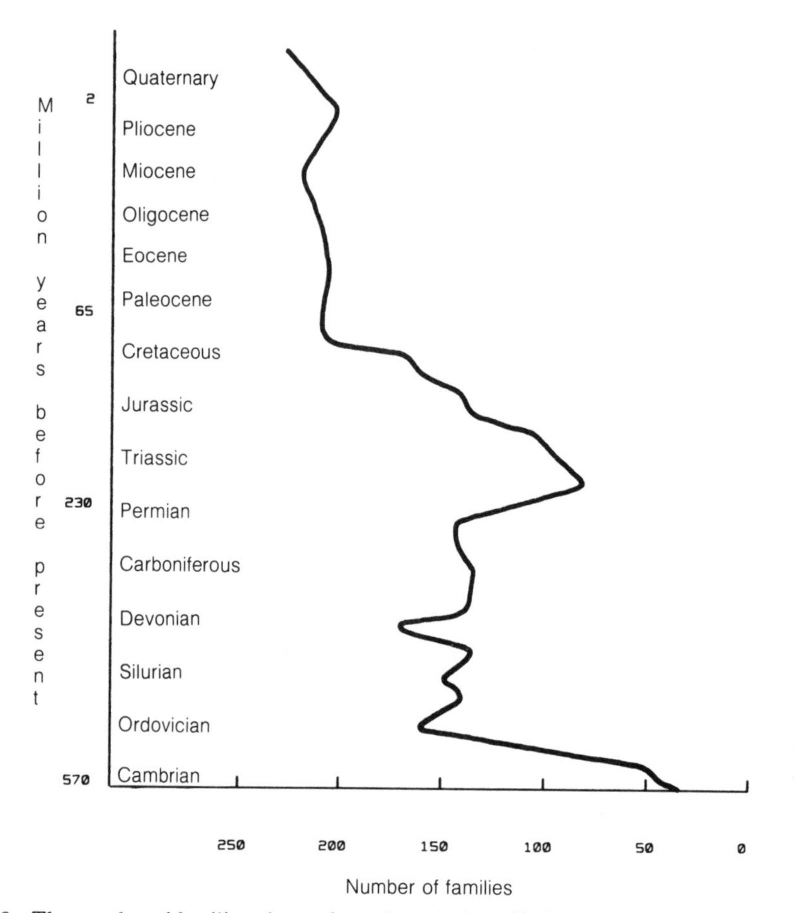

Fig. 5-2. The number of families of organisms through time. Notice the large drop at the end of the Permian.

Most, if not all, mass extinctions follow periods of global cooling, although not all climatic cooling has been accompanied by glaciation. Moreover, sea levels during many extinctions were no lower than they are now. However, sea levels are lowered significantly during periods of glaciation because large amounts of seawater are locked in expansive ice sheets. These sheets reduce the area of shallow water habitat, forcing crowded conditions and dwindling food stocks worldwide. The lowered temperatures also restrict the geographic distribution of species, which are confined to warmer regions around the equator.

Major extinction events also appear to be periodic, resulting from celestial influences, such as cosmic rays from supernovas or by large meteorite impacts (FIG. 5-3). There have been 10 or more large impacts over the last 600 million years, or about one roughly every 50 million years. Analysis of 13 major impacts spread over a period from 250 million years to five million years ago suggests a cratering rate of one per 28 million years, which is comparable to the 26-million-year extinction cycle (discussed in more detail in chapter six).

(COURTESY OF USGS)

Fig. 5-3. Meteor Crater in Arizona resulted when a large asteroid struck the Earth 22,000 years ago.

The extinctions might also be episodic with relatively long periods of stability followed by random, short-lived extinction events that only appear to be periodic. Most extinction episodes seem to eliminate certain types of species, and a comparison between the victims and survivors might lead to the ultimate causes of the extinctions. Although the dinosaurs suffered extinction at the end of the Cretaceous, they might not have done anything wrong biologically. It has even been suggested that the dinosaurs had been around long enough, were a mistake in the first place, and thus deserved extinction.

When the dinosaurs became extinct, the mammals happened to be in the right place at the right time, in evolutionary terms, to take over the world. At first, the majority of mammals were small, nocturnal creatures with a limited range. Apparently, certain survival characteristics of the mammals made them capable of replacing the dinosaurs. One of these traits might have been their greater intelligence; although many dinosaur species are also thought to have been fairly intelligent. The mammals were also warm blooded, and so it seems were some dinosaur species.

These adaptations would appear to have given the mammals a decisive advantage during times of environmental stress, but for some reason, were no use to the dinosaurs. Perhaps the mammal's small size helped pull them through the extinction; however, not all dinosaurs were giants, and many were no larger than most mammals alive today. Maybe the mammals preyed on dinosaur eggs and therefore won the battle of evolution through attrition.

Whatever were the mammals' advantages, they were able to out-compete and out-populate the dinosaurs. Perhaps over a period of several million years, they were able to completely replace them. The fact that the dinosaurs were not the only ones to go and that 70 percent of all known species became extinct at the end of the Cretaceous indicates that something in the environment made them all unfit to survive. Yet for unknown reasons, the mammals were barely affected.

It appears that the placental mammals came through the extinction and flourished at the expense of the marsupial mammals, which were hard hit. This evidence suggests that the more advanced placental reproduction, in which the embryo develops in the uterus, was a major reason for the mammals' great success. Yet, it is thought that some dinosaur species also gave live birth. Moreover, egg-laying fish, amphibians, reptiles, and birds still enjoy a high degree of reproductive success, as is indicated by the large populations of these animals throughout the world.

GRADUAL EXTINCTIONS

Catastrophic extinction events occur almost instantaneously in the geologic record. However, a resolution of several thousands of years over millions of years of geologic time is not possible. It therefore seems more likely that the extinctions of species occurred over lengthy periods, perhaps over a million years or more. Because of erosion or nondeposition of sedimentary strata, which preserves species for the fossil record, the die-outs only appear to be sudden.

Moreover, the fossil record can be extremely insensitive to major changes in groups of organisms. A family containing 60 species could be devastated so that only one species survives, and yet no record of the change would be recorded. At the end of the Cretaceous, the falling sea levels reduced sedimentation rates as well as fossilized species. Thus, what appears to have been a sudden break might indeed have occurred over an extended period of time.

From the time life first appeared on Earth, there has always been a gradual die-out of species called, *background* or *normal extinctions.* Major extinction events are separated by periods of lower extinction rates, and the difference between the two is only a matter of degree. There is also a qualitative as well as a quantitative distinction between background and major extinctions, and species regularly disappear—even during optimum conditions.

However, mass extinctions are not simply intensifications of the processes that operate during background times. In other words, the survival characteristics that develop during normal times become irrelevant during mass extinctions. This information suggests that mass extinction might be less discriminatory, with respect to the environment, than normal extinction and that different processes might be operating during times of mass extinction than those operating during normal times. Furthermore, the same types of species that succumb to mass extinctions also succumb to background extinction, only much fewer species die out.

Those species that survive mass extinctions are particularly hardy and resistant to subsequent random changes. They tend to occupy large geographical ranges that contain a large number of groups of related species (FIG. 5-4). Because a species survives extinction, however, it is not necessarily superior or better-suited to its environment. Those that become extinct might have been developing certain unfavorable traits during background times. This could occur even within the same organisms, as daughter species develop better survival skills and replace their parent species.

Fig. 5-4. The distribution of shelf faunas showing the geographical ranges of species.

Although the dinosaurs were keenly adapted to their environment, which accounts for their great success in dominating the entire world for 140 million years, a sudden change in environmental conditions might have done them in because they were incapable of quick adaptation. In other words, they were biologically suited for the conditions of the Mesozoic era, but apparently not to those of the following Cenozoic era.

FOSSIL PRESERVATION

The distinction between background and mass extinctions might be distorted by ambiguities in the fossil record, particularly when some species are favored over others for fossilization. Under certain geological conditions, which result in rapid burial without predation or decomposition (FIG. 5-5), the bodies of dead organisms are preserved to withstand the rigors of time.

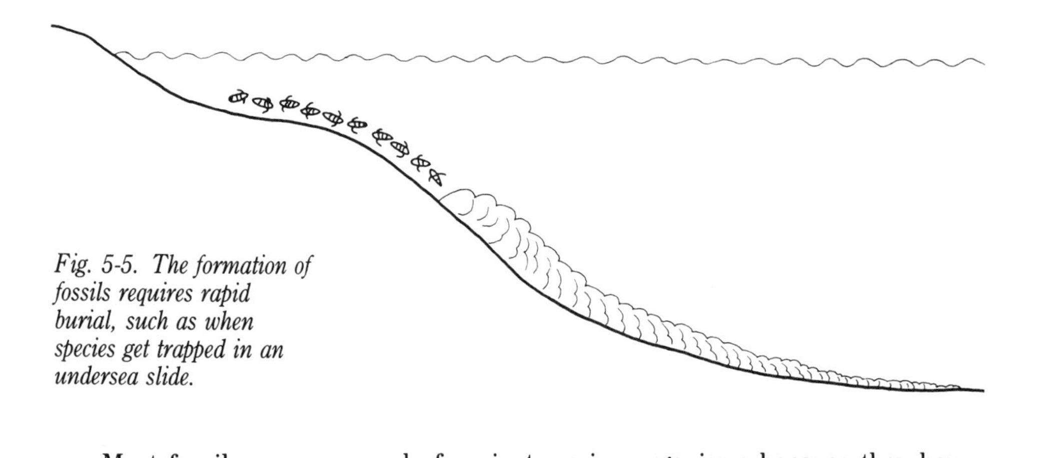

Fig. 5-5. The formation of fossils requires rapid burial, such as when species get trapped in an undersea slide.

Most fossils are composed of ancient marine organisms because they have been around the longest and were the most abundant. The ocean is also where the process of sedimentation takes place. Tiny shelled organisms provide a rain of calcite onto the ocean floor, which eventually hardens into limestone (FIG. 5-6), and captures the fossils of other organisms. Thus, the ocean provides an ideal setting for the preservation of marine species, and these organisms stand a better chance of becoming fossilized.

Some organisms, especially those with soft bodies like the late Precambrian fauna that lived prior to the arrival of shelled organisms, had a difficult time entering the fossil record. This is why fossils are dominated by organisms with hard skeletal remains. Therefore, shells, bones, teeth, and wood predominate the record of past life. So, fossils record a somewhat lopsided view of previous life on Earth.

Thus, only a small fraction of all the organisms that have ever lived were preserved as fossils. Normally, the remains of a plant or animal are destroyed through predation and decay. Although it seems that fossilization is easy for some organisms, for others it is nearly impossible. For the most part, the remains of organisms are recycled in the Earth, which is fortunate because otherwise the biosphere would soon become depleted of essential nutrients. Also, only a small fraction of all the fossils is exposed on the Earth's surface, and most of these are destroyed by weathering processes. This erosion makes for an incomplete fossil record, leaving many species unrepresented.

Generally, in order to become preserved in the fossil record, organisms must possess hard body parts, such as shells or bones (FIG. 5-7). Soft fleshy structures are quickly destroyed by predators or decayed by bacteria. Even hard parts left on the surface for any length of time are destroyed. Therefore, organisms must be buried rapidly to escape destruction by the elements and to be protected from the agents of weathering.

Delicate organisms, such as insects, are difficult to preserve and consequently are quite rare in the fossil record. Not only do they require protection from decay, but they must not be subjected to any pressure that would crush them. Marine organisms make better candidates for burial than those living on the land because the ocean is typically the site of sedimentation, whereas the land is normally the site of erosion.

The bones of extinct animals are much rarer than their footprints, and many animals are known only by their tracks and trails. Even clear outlines of claws or nails, the shape of the foot pad, and the pattern of scales can be recognized. Much information about an animal's life-style can be determined by analyzing its footprints—for instance, its mode of locomotion, its gait and speed, and whether it was solitary or gathered in herds. Unfortunately, few high-quality fossil footprints exist, and most are partially destroyed during the sedimentary process that buries and preserves them.

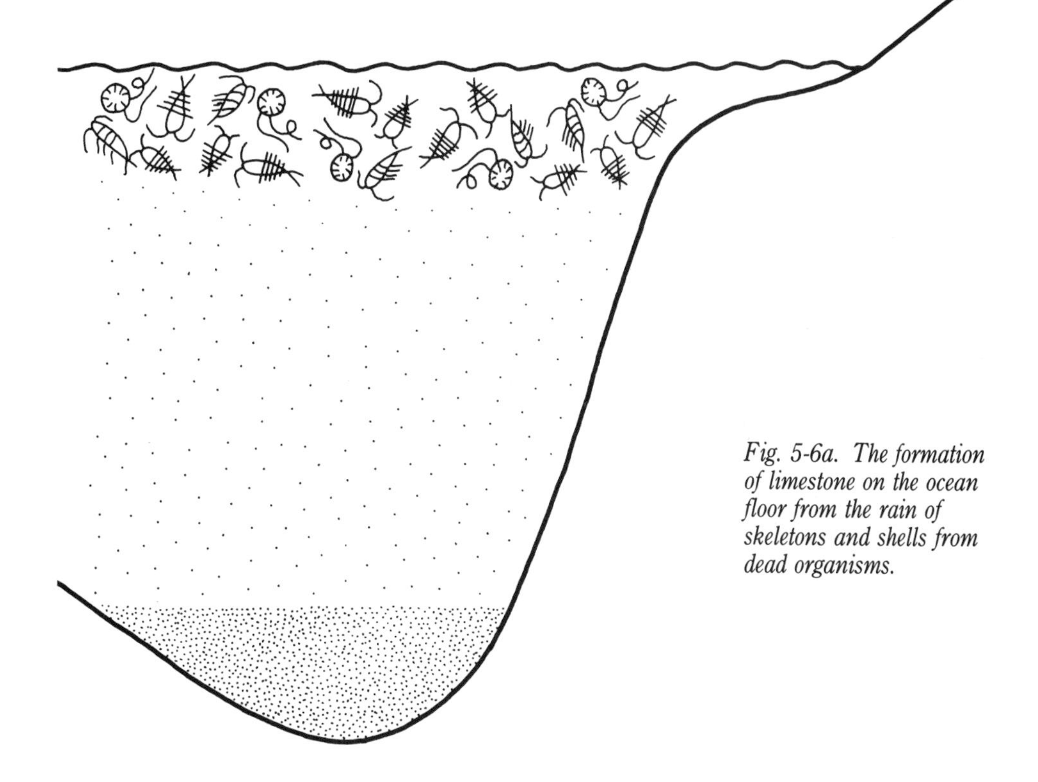

Fig. 5-6a. The formation of limestone on the ocean floor from the rain of skeletons and shells from dead organisms.

(PHOTO BY K. SEGERSTROM, COURTESY OF USGS)

Fig. 5-6b. Limestone of the Nantoco Formation in the Atacama Province Chile.

Animal tracks tell of the earliest invasion on dry land 370 million years ago. Primitive Devonian fish, similar to today's lungfish, crawled on their bellies from one pool to another, using their fins to push themselves along. The lobed-fin fish that first ventured on land gave rise to the four-legged amphibians; their tracks are found in formations from the late Devonian age. Amphibian footprints became abundant in the Carboniferous period, but they declined in the Permian, as a result of the rise of the reptiles.

(COURTESY OF NATIONAL PARK SERVICE)

Fig. 5-7. Restoration of dinosaur bones at Dinosaur National Monument.

Dinosaur tracks are the most spectacular of all fossil footprints (FIG. 5-8), and they are found in relative abundance in the terrestrial sediments of the Mesozoic age in many parts of the world. Some of these tracks are nearly mammal-like in their appearance, possibly indicating an intermediate species between reptile and mammal. By the end of the Mesozoic, dinosaur footprints disappeared entirely from the face of the Earth, and they were replaced by a multitude of mammal tracks.

RADIATION OF SPECIES

After a mass extinction occurs, the surviving species radiate outward to fill new environments. In turn, this extinction produces entirely new species. The new species might develop novel adaptations that give them a survival advantage over other species. The adaptations might lead to exotic-looking organisms that prosper during intervals of normal background extinction, but because of over-specialization, they are incapable of surviving mass extinction.

The geologic record seems to imply that nature is constantly experimenting with new forms of life; when one fails, it becomes extinct—never to be seen again. Once a species has become extinct, it is gone forever, and the odds of its particular combination of genes reappearing is infinitesimally small. Thus, evolution seems to run only in one direction, and although it perfects species to live at their optimum in their respective environments, it can never return to the past. Even though some future environment might perfectly match that of the warm Cretaceous period when the dinosaurs were prolific, they will never return.

(PHOTO BY R.J. DINGMAN, COURTESY OF USGS)

Fig. 5-8. Dinosaur trackway in Tarapaca Province, Chile.

The mammals and dinosaurs coexisted for more than 100 million years. After the dinosaurs became extinct, the mammals underwent an explosive evolutionary radiation (FIG. 5-9), which gave rise to many unusual species. The mammals successfully replaced the dinosaurs probably because of their more advanced biology, which placed them higher on the evolutionary scale. This biology might have given them a decisive edge during times of environmental stress, which allowed them to survive the end-Cretaceous extinction when the dinosaurs did not.

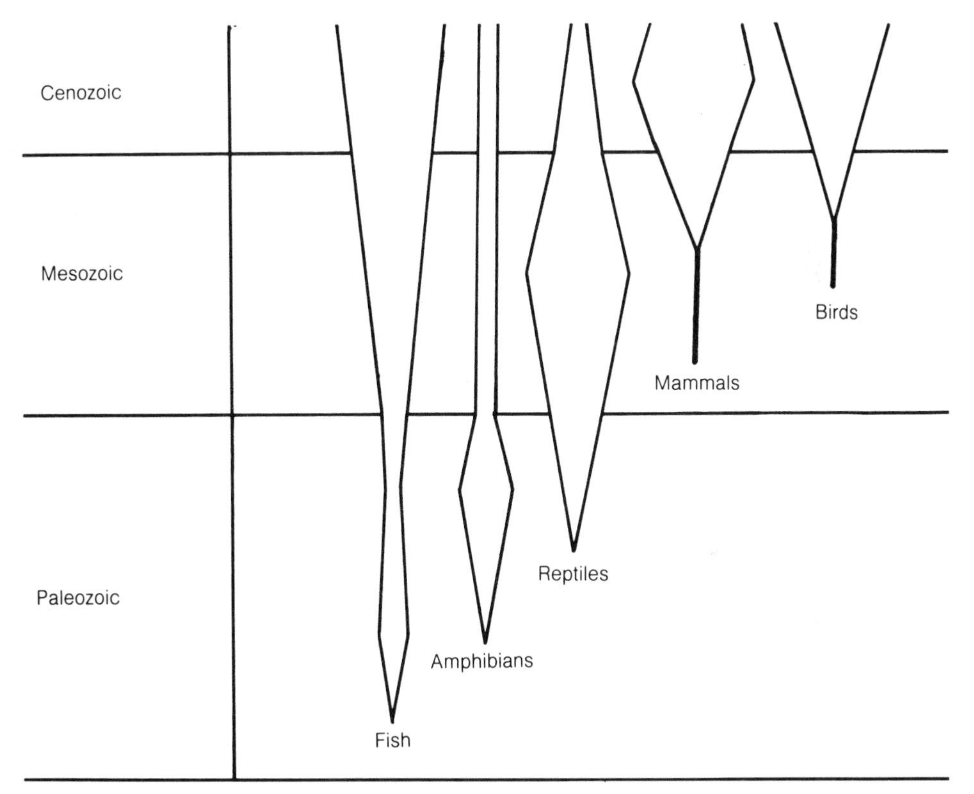

Fig. 5-9. The appearance and radiation of species.

Extremes in climate and topography during the Cenozoic period, brought on by rapid continental movements, provided a greater variety of living conditions than at any other equivalent span of Earth history. The rigorous environments presented many challenging opportunities for the mammals. As a response to these opportunities, species seem to show a burst of evolutionary development that gives rise to novel adaptations.

Higher life-forms, such as the mammals, evolve at higher rates than those lower on the evolutionary ladder, especially those living in the sea. Although the extinction in the oceans at the end of the Cretaceous was severe and many species died out, very few "radical" species evolved as a result. This is because ecological niches that were left vacant were simply occupied by closely related species. However, the situation was the opposite on land because when the dinosaurs (which represented the largest group of terrestrial animals) died out, the world was left open to invasion.

THE FINAL OUTCOME

A catastrophic event reduces the number of separate species, as well as the total number of species. Following this event, however, the biological system appears to be temporarily immune to random cataclysms. Those species that survive a mass extinction are particularly hardy and are resilient toward subsequent environmental changes. Moreover, after a major extinction event, fewer species are left to die out.

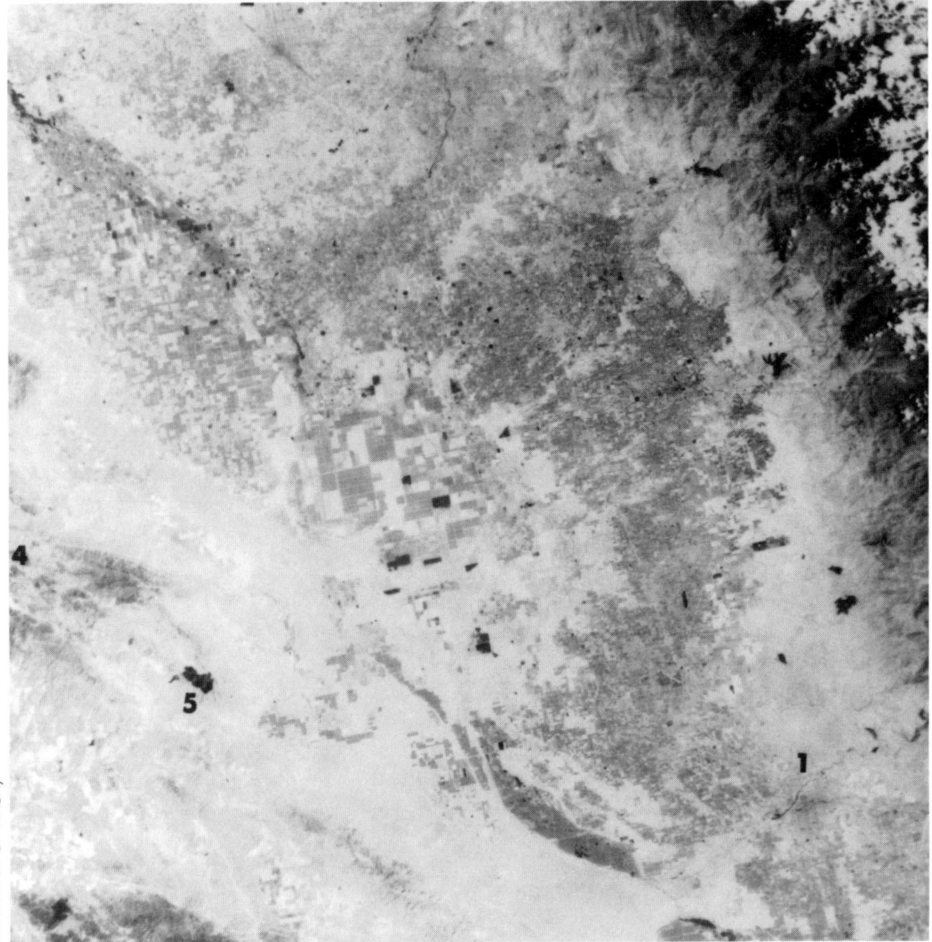

(COURTESY OF NASA)

Fig. 5-10. The San Juaquin Valley, California showing extensive agricultural activity.

Therefore, until many species have evolved, including extinction-prone types, any intervening catastrophes would have little effect. After each extinction, the biological world requires a recovery period before it is again ready to face another major extinction event. This is the reason why the mammals took so long to finally diversify after the death of the dinosaurs. Unfortunately for some mammal species, that diversification led to over-specialization, and they were knocked out of the running in the next extinction.

Each time a mass extinction occurs, the evolutionary clock is reset as though life was forced to start anew. Those species that survive radiate outward to fill entirely new niches, which subsequently produce entirely new species. A species is defined by its genetics and not by its *morphology* (body form). Also, the amount of genetic information carried in each cell rises steadily from simple to complex organisms. Moreover, different species might share certain similar physical attributes only because they share the same environment. For example, when reptiles and mammals returned to the sea, they acquired the appearance of fish.

After the great Permian extinction (240 million years ago), life experienced many remarkable advancements. The species that lived through the extinction event were similar to the populations of today. Many of these same species also survived the Cretaceous extinction, indicating they might have perfected survival characteristics that the other species lacked.

As the world recovered from the extinction, many regions of the ocean filled with numerous specialized organisms, and the overall diversity of species rose to unprecedented heights. But instead of developing entirely novel forms, such as those that began during the Cambrian explosion, the species that survived the extinction at the end of the Permian developed morphologies based on simple skeletal types with few experimental organisms.

If certain small carnivorous dinosaurs (with a brain-to-body-weight ratio characteristic of early mammals) had escaped extinction at the end of the Cretaceous, they conceivably could have continued to suppress the rise of the mammals. If they had done so, our own species might never have emerged to become the present masters of the Earth (FIG. 5-10).

6

Celestial Causes of Extinctions

GEOLOGISTS have devised a geologic time scale based on the appearance and disappearance of species. Some major breaks in geologic time occur at points of mass extinctions, when life is often forced to start anew. The mass extinctions also appear to be periodic, reoccurring about every 26 million years. Moreover, the extinction cycle is believed to be related to celestial phenomena. The gravitational attraction of an unseen sister star of the Sun might disturb comets, which surround the Solar System about a light-year away, and hurl them at Earth. The inevitable rain of objects that falls onto the Earth might therefore change the rules governing the evolution and extinction of life.

GALACTIC DUST CLOUDS

Since the great Permian extinction 240 million years ago, eight major extinction events have occurred. Some of the strongest peaks coincide with the boundaries between major geological periods. Extinction episodes seem to be periodic; they occur roughly every 26 million years (FIG. 6-1). Furthermore, longer intervals of 80 to 90 million years occur between major mass extinctions.

Exceptionally strong mass extinctions occur every 225 to 275 million years and might be linked to the Solar System's revolution around the Milky Way galaxy. If the Solar System entered a nebula or a galactic dust cloud on its journey around the galaxy, the dust could affect the Sun's output. It could also block solar energy

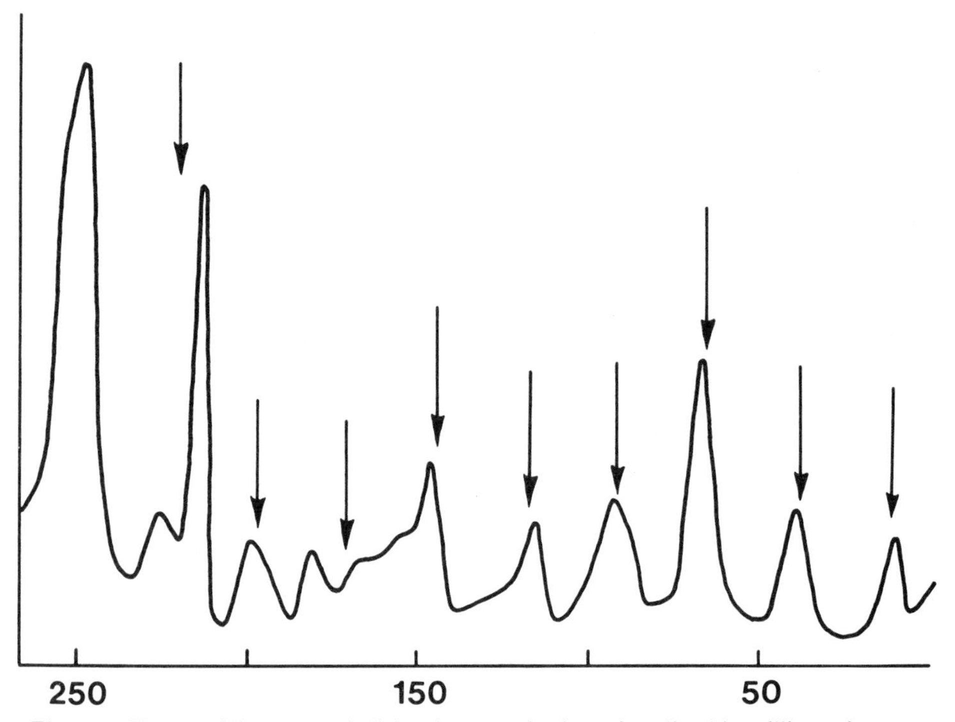

Fig. 6-1. The 26-million-year periodicity of mass extinctions. Ages listed in millions of years.

from reaching the Earth and cause climatic cooling, which in turn could have an adverse effect on the biosphere.

Major extinctions might not be periodic, but instead might reflect a clustering of several minor events at certain times, which as a result of the nature of the geologic record, only appear to be cyclical. In other words, random groupings of extinct species on a geologic time scale, which itself is uncertain, might be nothing more than a coincidence. Furthermore, a short period of rapid evolution might appear in the geologic record as though it was preceded by a mass extinction, when actually no extinction occurred.

Like all stars, the Sun oscillates up and down, perpendicular to the plane of the galaxy. It completes one cycle about every 60 to 80 million years, which is about the same length as one of the major extinction cycles and twice that of the 26- to 32-million-year cycle. Thus, the Sun crosses the midplane and reaches the maximum distance from it twice per cycle—roughly every 30 million years.

The Solar System's journey through the dense clouds located at the midplane of the galaxy could affect the Sun's output and reduce the Earth's *insolation*, the solar input. This reduction could in turn initiate climatic changes that would dramatically affect life on the planet. No evidence exists, however, to prove that the dust clouds are dense enough to block out the Sun during each passage through the midplane, a journey that could take upwards of several million years.

The Solar System is currently near the midplane of the galaxy and the Earth appears to be midway between major extinction events, the last two of which occurred approximately 37 million and 11 million years ago. This evidence might suggest that the extinction episodes instead coincide with when the Solar System reaches its furthest point from the galactic midplane. The extinction of the dinosaurs and large numbers of other species 65 million years ago occurred when the Solar System's distance from the midplane was nearly maximum.

When the Solar System reaches the upper or lower regions of the galaxy, it might be exposed to higher levels of cosmic radiation that originates from supernovas. The radiation could ionize the Earth's upper atmosphere and produce a haze that blocks out sunlight. Furthermore, if a giant star, like Betelgeuse which is 300 light years away, went supernova, the Earth could receive a blast of ultraviolet radiation and X-rays strong enough to burn off the ozone layer and destroy all life on the planet's surface.

SUPERNOVAS

On a cosmic time scale, supernovas appear quite frequently in our galaxy. However, we usually cannot see them because they are blocked by dark galactic dust clouds. When a giant star becomes a supernova, the massive nuclear explosion flings its outer layers into space at fantastic speeds. Large amounts of radiation are also released, producing deadly cosmic rays.

Toward the end of the last ice age, about 10,000 years ago, when large mammals such as the mastodons and mammoths (FIG. 6-2) became extinct, the supernova Vela is believed to have appeared. Such a supernova would emit a burst of gamma radiation and X-rays that possibly could destroy 80 percent of the Earth's ozone layer and allow harmful ultraviolet radiation from the Sun to penetrate the atmosphere. This radiation could have killed the vegetation that the large mammals grazed on and ultimately caused their own demise.

Cosmic rays are the most energetic radiation known—far more powerful than anything physicists have created in the laboratory. The bombardment of nitrogen atoms in the upper atmosphere by cosmic rays creates radioactive carbon-14 (FIG. 6-3), which enters the tissues of all living things. This radioactivity enables scientists to accurately date events of the past using radiocarbon techniques. Cosmic rays were first discovered by the Austrian physicist Victor Hess in 1912. Hess made a high-altitude balloon flight with a radiation counter and discovered that the intensity of the cosmic rays increased with altitude, indicating that they originated outside the Earth.

The seemingly incessant rain of cosmic rays that bombard the Earth consists of atomic nuclei, protons, electrons, and gamma- and X-radiation. Most of the cosmic radiation that strikes the Earth today is believed to have originated from a supernova explosion that occurred about 10,000 years ago at a distance of 15 parsecs. That distance is roughly 10 times farther than the nearest star, Proxima Centauri, which is about four light-years from Earth.

When a giant star goes supernova, the force of the explosion crushes the star's core, and compresses it into a neutron star—about the size of the Earth. The

Fig. 6-2. The mammoth (top) and the mastodon (bottom) went extinct at the end of the last ice age.

expanding shell of the supernova injects huge amounts of gas and debris into the galaxy. The Solar System might enter such a dust cloud every hundred million years or so. If the Solar System happened to pass through relatively dense regions of the dust cloud, the material falling into the Sun might affect its output. The dust might also directly alter the amount of sunlight the Earth receives by blocking the Sun.

The passage through the dust cloud could take several million years and might have been responsible for some of the earlier episodes of glaciation, the first of which appears to have occurred no earlier than about 2.7 billion years ago.

It also seems doubtful that passing through a dusty arm of the galaxy would have caused the continuous temperature decrease from the Cretaceous period to the present. However, a cloud of interstellar gas does appear to be streaming through the Solar System at a speed of 50,000 miles per hour.

During its lifetime, the Solar System has passed through at least 100 dense gas clouds. As the Solar System traverses through the interstellar cloud, a journey that can take 100,000 years, the Earth could acquire large amounts of molecular hydrogen. The hydrogen could react with chemical constituents in the upper

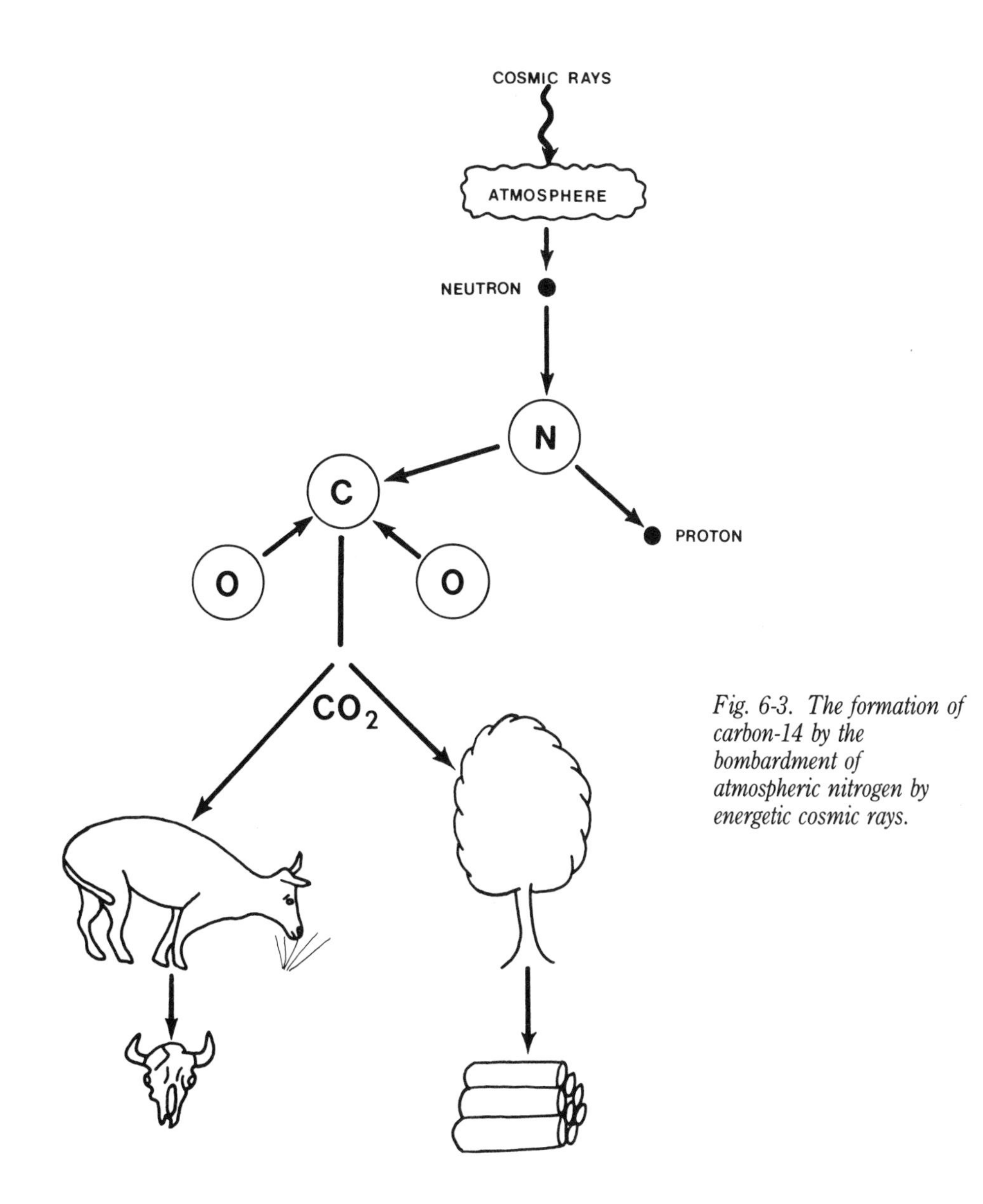

Fig. 6-3. The formation of carbon-14 by the bombardment of atmospheric nitrogen by energetic cosmic rays.

atmosphere to produce water vapor, which would condense into clouds. The clouds would reflect solar radiation and lower surface temperatures several degrees. If this action was sustained for several thousand years, it could initiate an ice age. The resulting colder climate might cause a mass extinction.

SOLAR ACTIVITY

For centuries, astronomers have referred to a property of the Sun as the *solar constant.* It is believed that the total amount of solar energy impinging on the Earth has remained fairly steady for the last 2 billion years. Prior to this time, the Sun became progressively more luminous. The solar constant depends on the Sun's luminosity and the Earth's orbit. The Sun's luminosity depends on its size and surface temperature. A reduction of the solar constant by just a few percent is sufficient to engulf the Earth in ice. On the other hand, an increase of only a few percent would make this planet a desert, unbearably hot and uninhabitable.

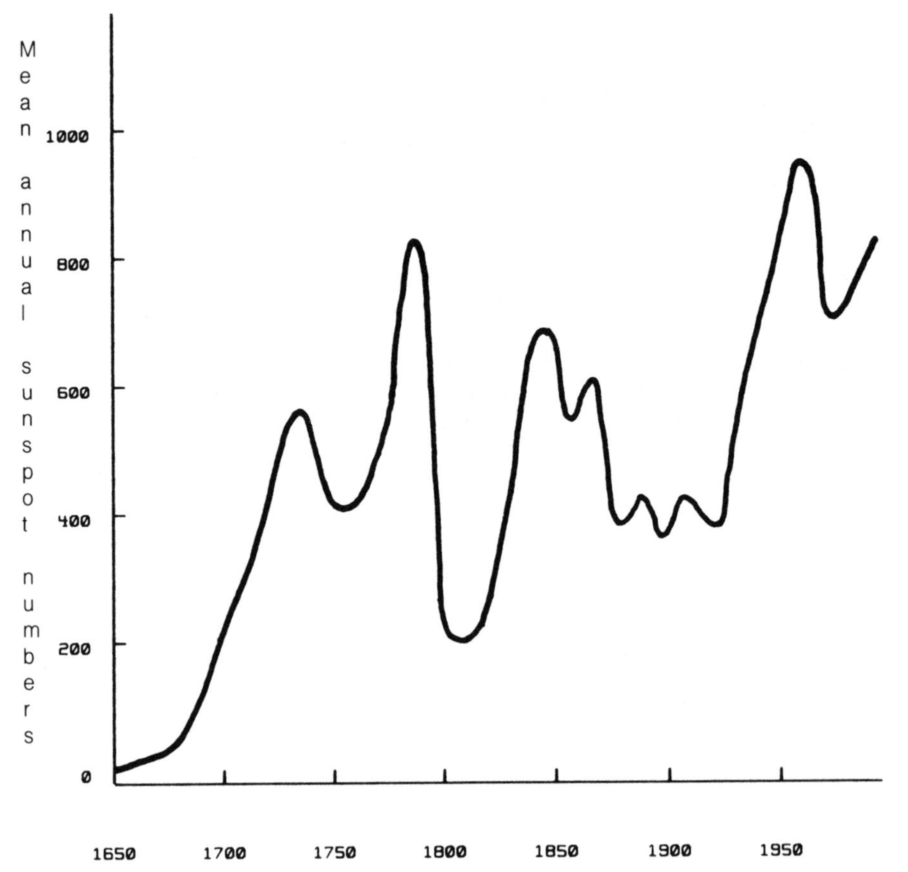

Fig. 6-4a. Sunspot activity through time. The large dip in the late 1600s occurred during the Little Ice Age.

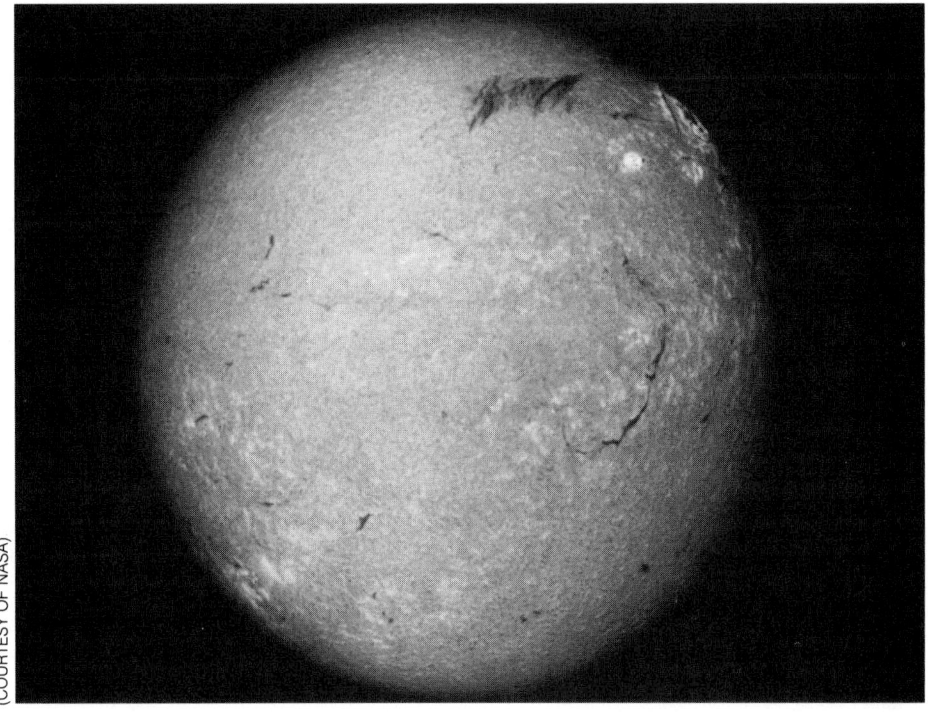

(COURTESY OF NASA)

Fig. 6-4b. Solar storms are responsible for activities on Earth, including magnetic and electrical disturbances and can even affect the weather.

The average solar output of the Sun seems fairly steady over the short run, although its output has fluctuated slightly during periods of sunspot activity (FIG. 6-4). However, during its entire lifetime, the Sun's luminosity has been steadily increasing. About four billion years ago, the Sun was about 8.5 percent smaller and its luminosity was three to four percent less than it is today. That means the solar constant was as much as 30 percent less than it is today. Thus, the Earth's equator was then receiving only as much solar energy as Antarctica does now.

Over the next four billion years, the Sun will not get much hotter. It will, however, continue to expand as it depletes the hydrogen in the core and begins to consume that which is in the outer layers. The solar luminosity is controlled by the thermonuclear reactions in the core and the heat transferring properties of the gases in the outer layers (FIG. 6-5). The conversion of hydrogen into helium for thermonuclear energy constantly depletes the hydrogen and pollutes the core with helium, which subsequently increases the Sun's luminosity over a very lengthy time period. In about five billion years, the Sun will balloon into a red giant to the orbit of Mercury; the intense heat will incinerate the Earth.

The reactions in the Sun apparently do not follow prescribed nuclear physics. This statement is demonstrated by the so-called "neutrino problem." Since 1968,

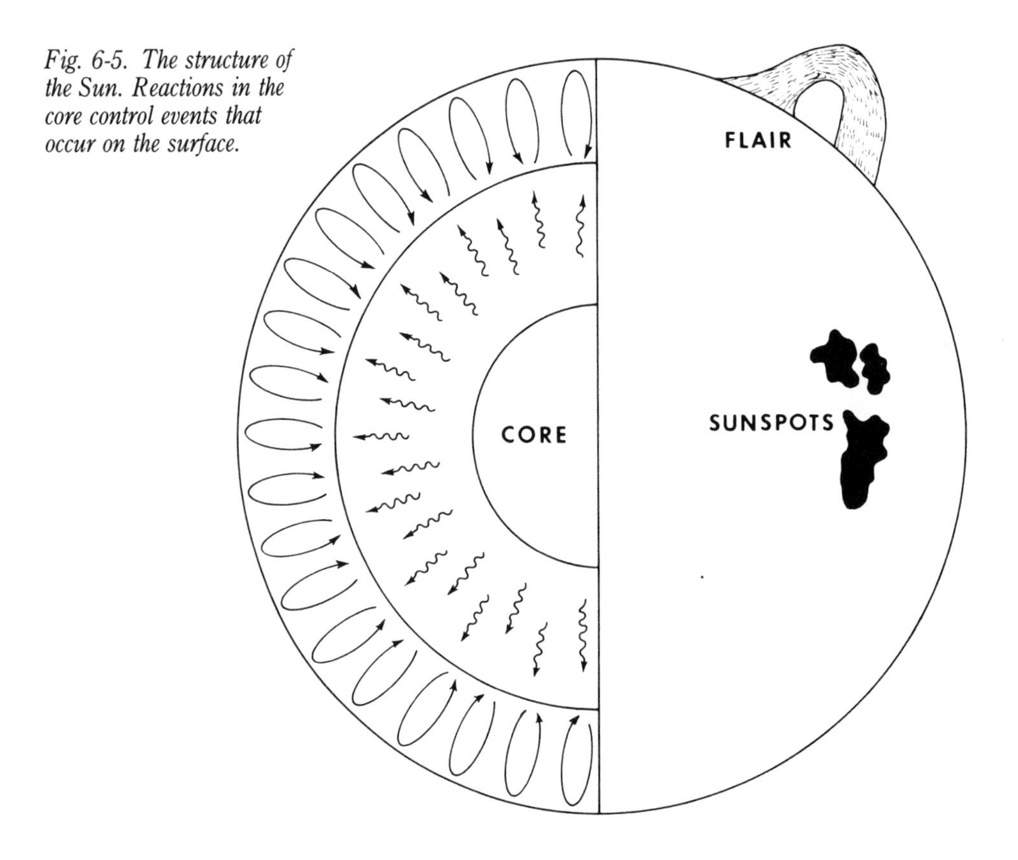

Fig. 6-5. The structure of the Sun. Reactions in the core control events that occur on the surface.

scientists have been trying to catch neutrinos, which are tiny particles produced by the nuclear reactions in the Sun. They buried a huge vat of cleaning fluid surrounded by sensitive electronic detectors deep in a South Dakota gold mine. However, they have been able to detect only about 30 percent of the neutrinos that should have been coming from the Sun.

It is postulated that the Sun's core temperature, estimated at about 15 million degrees Celsius, is less than the theoretical amount. This possibility might be responsible for the production of fewer neutrinos. A lowered core temperature, however, would give the Sun a different luminosity than what is observed. Moreover, if the core was actually cooling down, the Sun's output would diminish considerably, and all life on Earth would be in real trouble.

NEMESIS AND PLANET X

Surrounding the Sun about a light-year away lies a shell of over a trillion comets with a combined mass of 25 Earths. This conglomeration of comets is called the *Oort Cloud,* named for the Dutch astronomer Jan Kendrick Oort, who first postulated its existence. Comets are leftovers from the formation of the Solar System and are composed of an inner core of rocky material surrounded by ice.

Comets are as large as tens of miles in diameter and they have highly elliptical orbits that can take them within the inner Solar System (FIG. 6-6).

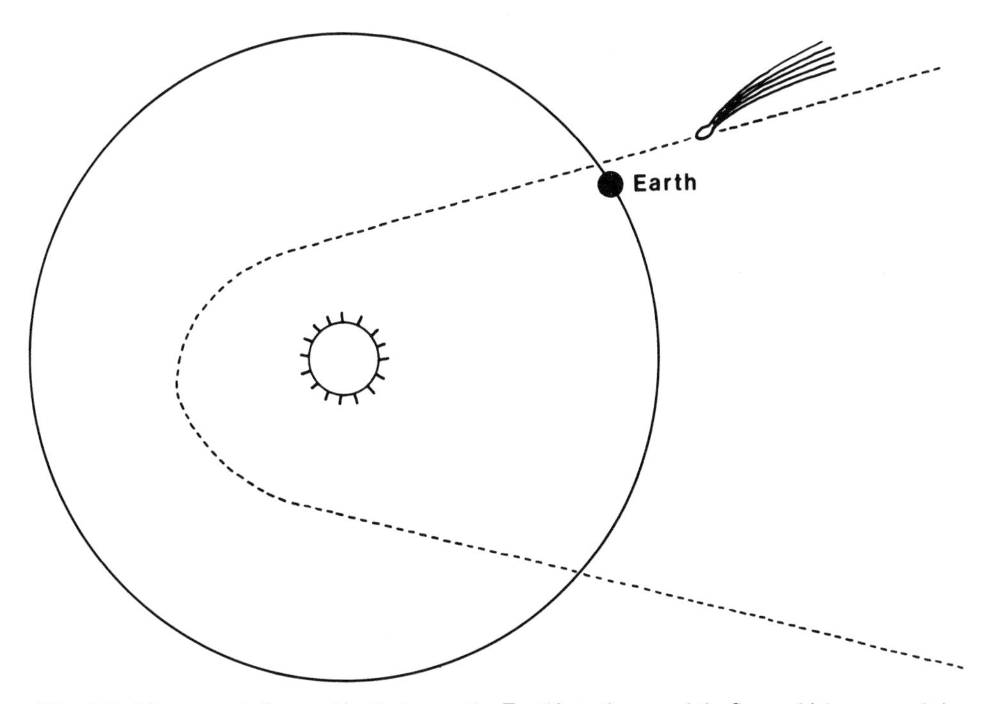

Fig. 6-6. Many comets have orbits that cross the Earth's path around the Sun, which can result in a collision.

If the Earth happens to obstruct one or more of these icy visitors from outer space, the collision could be devastating to life on this planet, and cause almost instantaneous extinctions. A massive comet shower that would involve perhaps thousands of comets impacting all over the Earth, might also help explain the disappearance of the dinosaurs and other species at the end of the Cretaceous.

The comets would shock heat the atmosphere as they streak toward the surface, combining nitrogen with oxygen to form nitric oxide. This compound would react with atmospheric water vapor to form a strong solution of nitric acid. As a result, a strong acid would rain down on the Earth. A massive die-out of species could result because most organisms cannot tolerate high acid levels in their environment. Nitric oxide also destroys ozone, and the erosion of the ozone layer by a comet shower could leave all inhabitants on Earth exposed to the Sun's deadly ultraviolet radiation.

One popular theory for sudden mass extinctions that seem to reoccur roughly every 26 million years deals with a hypothetical companion star of the Sun, named *Nemesis,* in honor of the Greek goddess who was thought to dish out punishment on Earth. Nemesis is believed to revolve around the Sun in a highly elliptical orbit that inclines steeply against the plane of the Solar System. So, it would be possible for Nemesis to approach the Oort Cloud every 26 million years. Its strong gravitational pull would distort the orbits of the comets in its vicinity and drive a swarm of them toward the inner Solar System.

Some of the comets might rain down on the Earth, and the damage could force the extinction of large numbers of species. The periodicity of the extinctions might also be attributed to the Earth's movement through the galactic midplane, where thick gas and dust clouds produce gravity anomalies strong enough to break loose comets in the Oort Cloud and launch them toward Earth.

Somewhere in the depths of space, too dim to be seen by the most powerful telescopes, is thought to be a tenth planet, named *Planet X.* It is believed to lie outside the orbit of Pluto, perhaps about 10 billion miles from the Sun. It probably revolves around the Sun in an elongated orbit that tilts steeply against the orbits of the other planets, taking possibly one thousand years to complete one revolution. Planet X would have to be less than five times the mass of the Earth because a larger planet probably would have been detected by now.

Astronomers think they might have detected the presence of Planet X indirectly by its gravitational influence on Uranus and Neptune. These planets were deflected from their orbits during the last century, but for some reason not in this century. This evidence seems to indicate that Planet X must be in a rather peculiar orbit. Thus, ever 28 million years, Planet X crosses a disc or belt of comets that are thought to lie in the plane of the Solar System beyond the orbit of Neptune. The gravity of Planet X would disrupt the orbits of some comets and send them crashing down on the Earth, disrupting all life in the process.

ASTEROID COLLISIONS

Asteroids are often called flying mountains (FIG. 6-7). They range in size from about one mile to hundreds of miles wide. Asteroids have produced more than 120 known craters throughout the world (TABLE 6-1). Up to three asteroid collisions, producing craters over 10 miles wide, are expected to occur somewhere on Earth every million years.

No one knows for sure how these large rock fragments fall into orbits that cross our planet's path. Apparently, they run in nearly circular orbits around the Sun for as long as a million years or more. Then, for unknown reasons, possibly as a result of a passing comet or the gravitational pull of Jupiter, their orbits suddenly stretch and become so elliptical that some actually collide with the Earth and the Moon. The craters on the Moon (FIG. 6-8) are much more apparent and numerous because of the lack of weathering processes that have destroyed most craters on Earth.

When a large asteroid slams into the Earth, the huge explosion tosses up a massive amount of sediment and excavates a deep crater. The finer material is lofted high into the atmosphere, where it shades the Earth and lowers global temperatures. Acids produced by a large number of meteors or comets that enter the atmosphere could also upset the ecological balance by introducing acid rain into the environment.

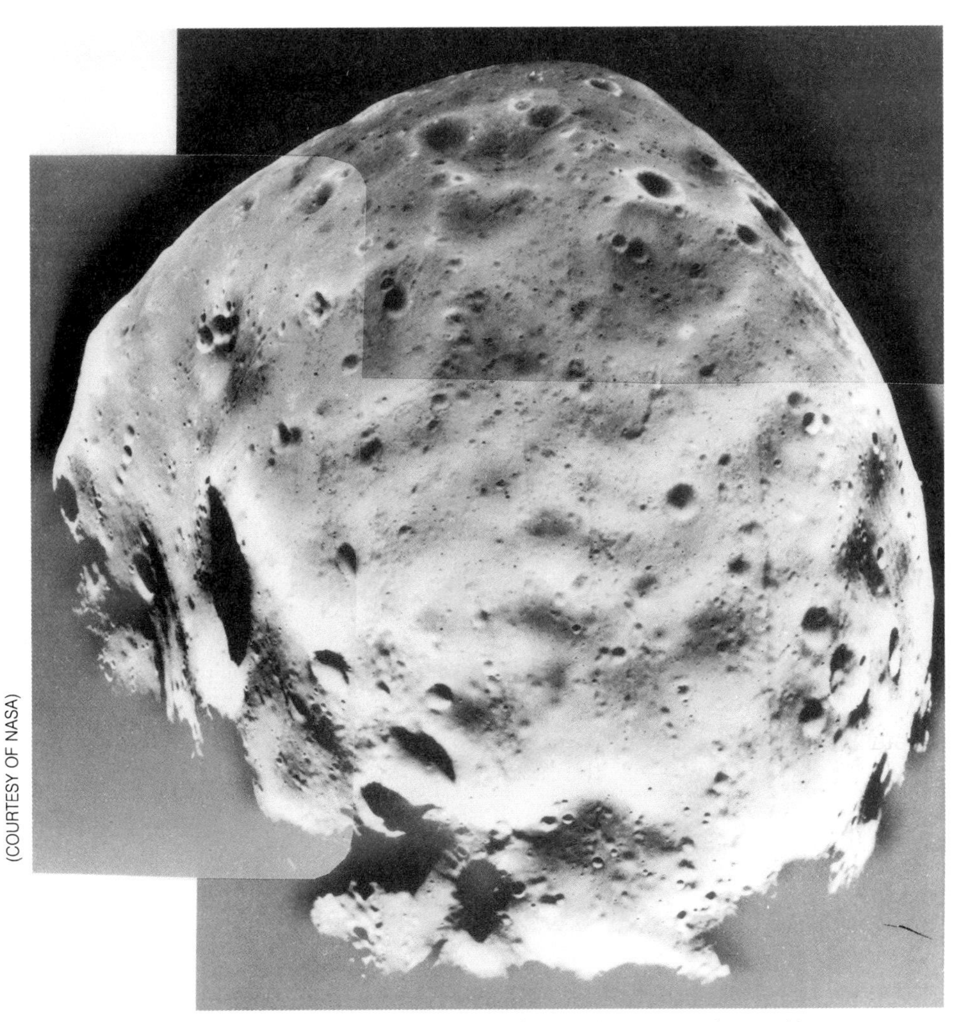

Fig. 6-7. Mars' moon Phobos is believed to be a captured asteroid.

The dinosaurs originated after the great Triassic extinction 210 million years ago, when a large asteroid is believed to have impacted in Quebec, creating the 60-mile-wide Manicouagan reservoir (FIG. 6-9). The impact might have caused the extinction of nearly half of the ancient reptile families and paved the way for the rise of the dinosaurs. When the Cretaceous ended, the Earth might have again been struck by large asteroid—with the explosive force of 100 trillion tons of TNT or about one million eruptions of Mount St. Helens. Thus, the dinosaurs might have been both "created" and destroyed as a result of asteroid collisions.

A massive bombardment of meteors or comets might also strip away the Earth's ozone layer in the upper atmosphere, and leave all species on the surface vulnerable to the Sun's deadly ultraviolet rays. This radiation would kill land

TABLE 6-1. Location of Major Meteorite Impact Structures

NAME	LOCATION	DIAMETER IN FEET
Al Umchaimin	Iraq	10,500
Amak	Aleution Islands	
Amguid	Sahara Desert	
Aouelloul	Western Sahara Desert	825
Bagdad	Iraq	650
Boxhole	Central Australia	500
Brent	Ontario, Canada	12,000
Campo del Cielo	Argentina	200
Chubb	Ungava, Canada	11,000
Crooked Creek	Missouri, USA	
Dalgaranga	Western Australia	250
Deep Bay	Saskatchewan, Canada	45,000
Dzioua	Sahara Desert	
Duckwater	Nevada, USA	250
Flynn Creek	Tennessee, USA	10,000
Gulf of St. Lawrence	Canada	
Hagensfjord	Greenland	
Haviland	Kansas, USA	60
Henbury	Central Australia	650
Holleford	Ontario, Canada	8,000
Kaalijarv	Estonia, USSR	300
Kentland Dome	Indiana, USA	3,000
Kofels	Austria	13,000
Lake Bosumtwi	Ghana	33,000
Manicouagan Reservoir	Quebec, Canada	200,000
Merewether	Labrador, Canada	500
Meteor Crater	Arizona, USA	4,000

TABLE 6-1. *Continued*

NAME	LOCATION	DIAMETER IN FEET
Montagne Noire	France	
Mount Doreen	Central Australia	2,000
Murgab	Tadjikstan, USSR	250
New Quebec	Quebec, Canada	11,000
Nordlinger Ries	Germany	82,500
Odessa	Texas, USA	500
Pretoria Saltpan	South Africa	3,000
Serpent Mound	Ohio, USA	21,000
Sierra Madera	Texas, USA	6,500
Sikhote-Alin	Sibera, USSR	100
Steinheim	Germany	8,250
Talemzane	Algeria	6,000
Tenoumer	Western Sahara Desert	6,000
Vredefort	South Africa	130,000
Wells Creek	Tennessee, USA	16,000
Wolf Creek	Western Australia	3,000

plants and animals and the primary producers in the surface waters of the ocean. Indeed, *plankton*, the small floating plants and animals of the sea, has the highest rate of extinction of any group of marine organisms. Within 500 thousand years after the Cretaceous closed, 90 percent of the species of plankton had disappeared.

A collision with a large extraterrestrial body could result in almost instantaneous extinctions. A large asteroid could explode with a force one thousand times greater than all of the nuclear weapons in the world. This explosion would send 500 billion tons of sediment into the atmosphere. Such an impact would also produce a deep crater that could reach the molten rocks beneath the crust and produce a massive volcanic eruption. Besides generating substantial quantities of dust by the impact itself, huge quantities of volcanic ash would be injected into the atmosphere and choke out the Sun.

The heat produced by the compression of the atmosphere and impact friction could set global-wide forest fires. The fires would probably consume 80 percent of the surface biomass, and turn the Earth into a smoldering cinder. This fire would destroy most terrestrial habitats and cause extinctions of massive proportions. A heavy blanket of dust and soot would cover the entire globe and linger for months. This cover would not only cool the Earth, but it would also halt photosynthesis, and kill species in tragic numbers. The species in the tropics, which require warmth and sunshine, would be especially hard hit.

In the aftermath of such an impact, would be a year of darkness under a thick brown smog of nitrogen oxide. Waters would be poisoned by trace metals leached out of the soil and rock, and global rains would be as corrosive as battery acid. Plants, which could have survived in the form of seeds and roots, would be relatively unscathed. As a result of the high acidity levels, marine organisms with calcium carbonate shells would dissolve, but organisms with silica shells (such as diatoms) would do relatively well. Land animals living in burrows and creatures living in lakes buffered against the acid would generally survive the impact.

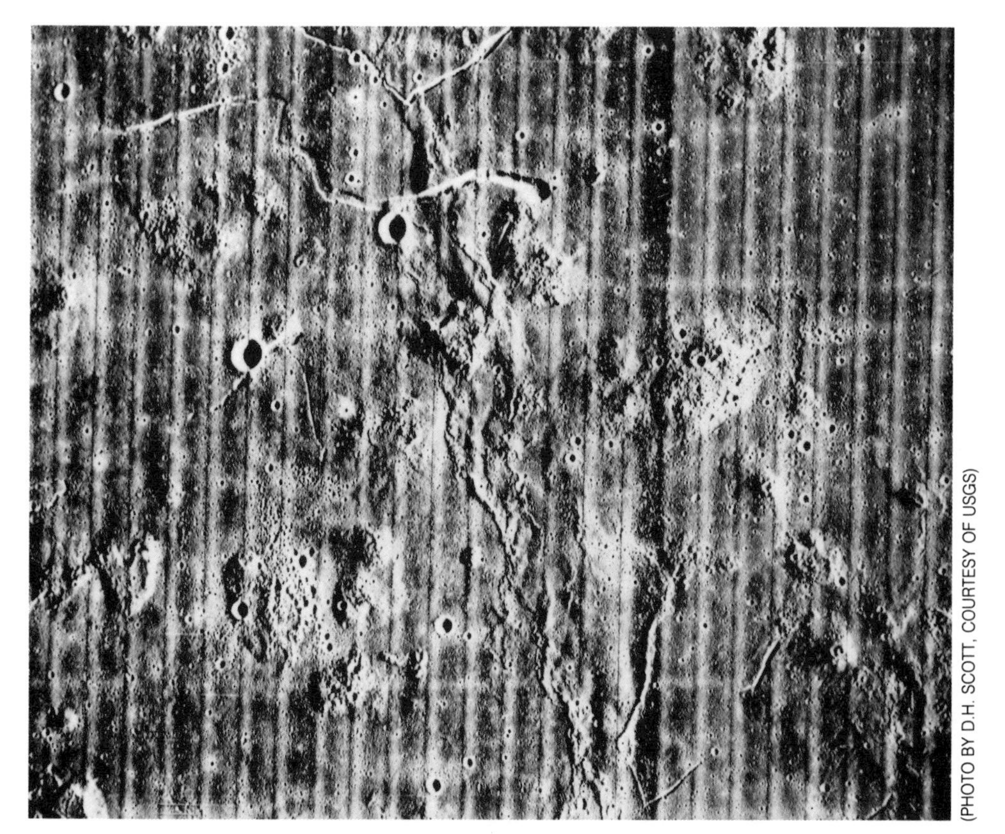

(PHOTO BY D.H. SCOTT, COURTESY OF USGS)

Fig. 6-8. The heavily cratered Marius Hills region on the Moon.

(COURTESY OF NASA)

Fig. 6-9. The Manicouagan reservoir outlines a 60-mile-wide impact structure in Quebec, Canada.

It is postulated that 10 or more major asteroids have collided with Earth in the last 600 million years. Moreover, the supposed impact of a 6-mile-wide asteroid at the end of the Cretaceous appears to be unlike any other. The boundary rocks between the Cretaceous and the Tertiary periods throughout the world (FIG. 6-10) contain a thin layer of fallout material. This mud contains shock-impact sediments, spherules (small, glassy beads), organic carbon from forest fires, a mineral called *stishovite* (found only at impact sites), meteoric amino acids, and an unusually high iridium content. Moreover, the iridium is not uniform throughout the world because microbes can either enhance or erase the iridium concentration in rock. Bacteria causes iridium to enter solution, suggesting that microbes could have erased part of the original iridium layer or spread it to deeper layers.

It is still fiercely debated among scientists whether the iridium came from an asteroid impact or from massive volcanic eruptions, which could also be a major source of iridium. However, volcanoes do not produce the type of shock-impact features on sediment grains—like those found at impact sites. The spherules at the end-Cretaceous boundary appear to have been created by impact melt and not by volcanism. The stishovite must have originated from an impact because the mineral breaks down at about 300 degrees Celsius, far below the temperatures generated by volcanoes. However, it still remains unexplained how the meteoric amino acids escaped destruction from the heat produced by the impact or from ultraviolet rays after the acids settled on the Earth's surface.

(PHOTO BY R.W. BROWN, COURTESY OF USGS)

Fig. 6-10. The Cretaceous-Tertiary boundary rocks—shown at the base of the dark streak, just above the white sandstone in the center—near Rock Springs, Wyoming.

The geologic record holds clues to other giant impacts associated with iridium anomalies that also happen to coincide with extinction episodes. However, these are not nearly as strong as the iridium concentrations at the end of the Cretaceous, which are as much as one thousand times greater than the background levels. This concentration suggests that the end-Cretaceous event might have been unique in the history of life on Earth.

7

Terrestrial Causes of Extinctions

CYCLES of terrestrial phenomena might best explain the apparent periodicity of mass extinctions. The *geologic* or *rock cycle* involves the circulation of convection currents in the mantle, which controls plate tectonics and consequently all activities occurring on the surface of the Earth. The dance of the continents has been an ongoing process for at least the last 2.5 billion years, if not longer. The breakup and assembly of continents causes dramatic climate and sea-level changes. These cycles, however, occur only over hundreds of millions of years.

Magnetic field reversals also coincide with extinctions. They do not appear to be cyclic; however, they might be associated with other periodic phenomena, such as the impacts of large asteroids, which seem to account for about half of the reversals. Asteroid impacts could also trigger volcanoes that are poised to erupt. The effects of impact volcanism might explain the many characteristics of environmental crises at important geologic boundaries. The impacts could even initiate glaciation; although other forces, such as the Earth's orbital motions, appear to have a stronger influence. Nevertheless, these changes in the Earth have a major effect on all living things.

VOLCANIC ERUPTIONS

Over the past 250 million years, 11 distinct episodes of flood basalt volcanism have occurred. The large eruptions create a series of separate, but overlapping, lava flows. Many of these exposures worldwide have a terracelike appearance, known as *traps*—from the Swedish word for "stairs." Many flood basalts are located near continental margins, where great rifts separated the present continents from Pangaea. Others, like the Columbia River basalts of the northwestern United States, are related to hot-spot activity, whereby plumes of hot mantle rocks rise from great depths to the surface.

The episodes of flood basalt volcanism were relatively short-lived, with major phases lasting less than 3 million years. The timing of major outbreaks correlates well with the occurrence of mass extinctions of marine organisms. Furthermore, the episodes appear to be somewhat periodic, occurring about every 32 million years. During the eruption of a major basaltic lava flow, vigorous fire fountains (FIG. 7-1) inject a large mass of sulfur gas into the atmosphere, where they are converted into an acid that causes severe climatic and biological consequences.

(PHOTO BY G.A. SMATHERS, COURTESY OF NATIONAL PARK SERVICE)

Fig. 7-1. Fire fountains from an eruption of a Hawaiian volcano.

A large number of volcanoes erupting over a long time interval could lower global temperatures by injecting huge amounts of volcanic ash and dust into the upper atmosphere (FIG. 7-2). Heavy clouds of volcanic dust have a high albedo and reflect much of the solar radiation back into space, thereby shading the Earth and lowering global temperatures. This climatic change could cause the mass extinctions of plants and animals by reducing the rate of global photosynthesis.

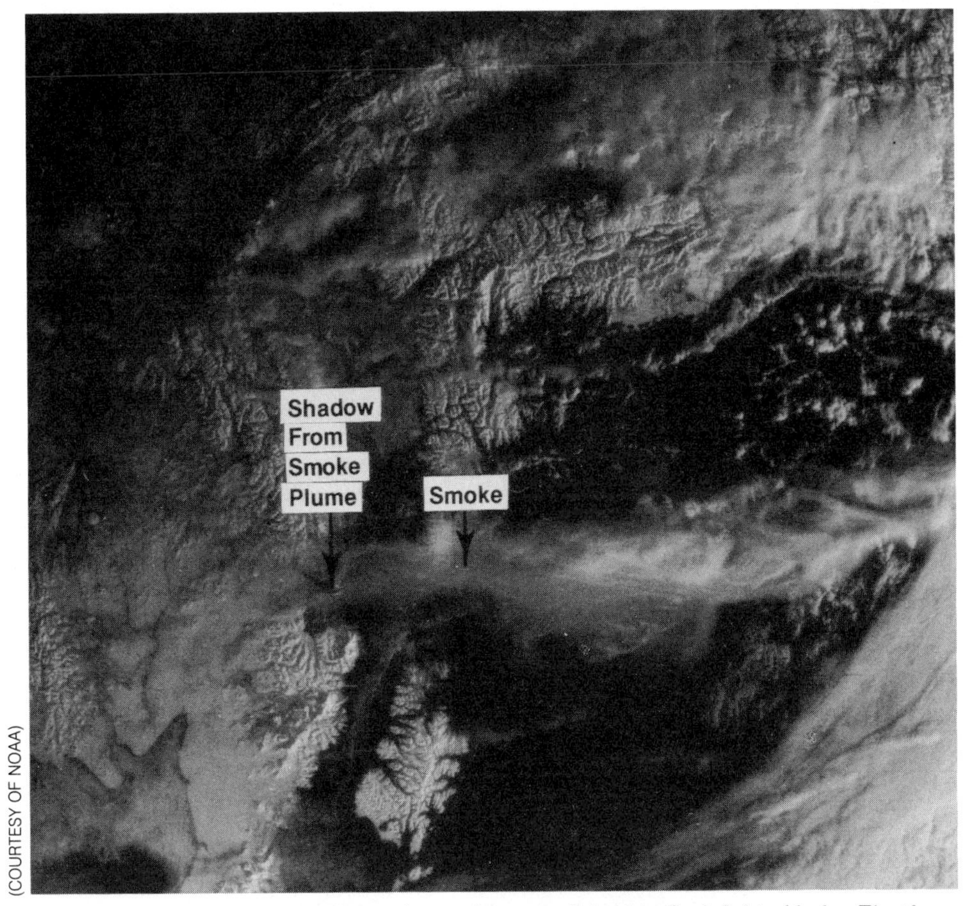

Fig. 7-2. The January 22, 1976 eruption on Augustine Island in Cook Inlet, Alaska. The plume of ash and smoke rose and then flowed southwestward over the Gulf of Alaska.

A five percent reduction in solar radiation reaching the Earth's surface could result in a drop in global temperatures by as much as 10 degrees Fahrenheit, enough to initiate an ice age. The long-term cooling would allow glaciers to expand and lower the sea level, which would limit marine habitat area. The lowered temperature could also adversely affect the geographical distribution of species and confine warmth-loving organisms to the narrow regions around the tropics.

Acid rain from extensive volcanic activity, 100 times as intense as that occurring worldwide today, could cause the widespread destruction of terrestrial and marine species by defoliating plants and altering the acid/alkaline balance in the ocean. Acidic gases spewed into the atmosphere might also deplete the ozone layer and allow deadly ultraviolet radiation from the Sun to scorch the planet.

Volcanoes directly affect the Earth's climate by altering the composition of the atmosphere. Large volcanic eruptions spew so much ash and aerosol gas into the atmosphere they block out sunlight. Volcanic dust also absorbs solar radiation and thereby heats the atmosphere, causing thermal imbalances and unstable climate conditions. Some scientists cite massive volcanic eruptions as the reason for the extinction of the dinosaurs.

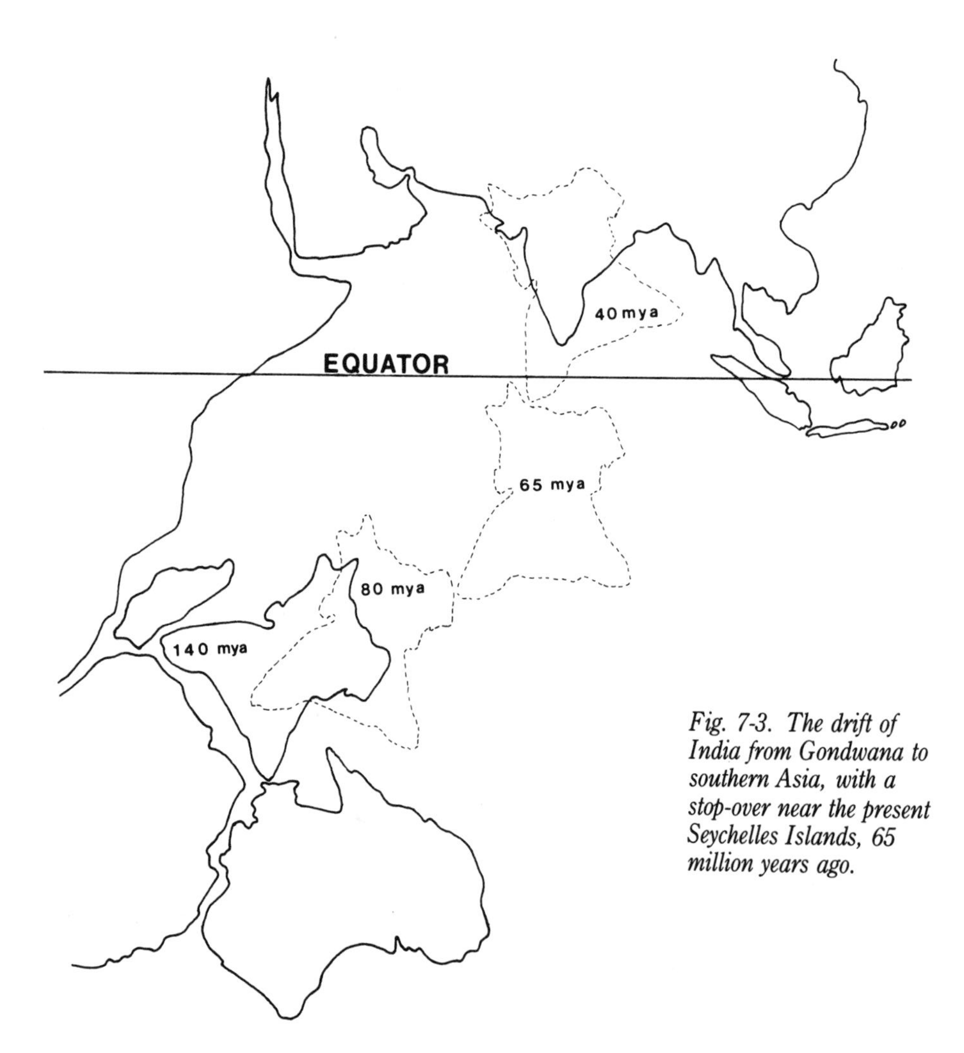

Fig. 7-3. The drift of India from Gondwana to southern Asia, with a stop-over near the present Seychelles Islands, 65 million years ago.

About 65 million years ago, at the end of the Cretaceous period, a giant rift split the west side of India, and huge volumes of molten lava poured onto the surface. Nearly 500 thousand square miles of lava were released in less than 500 thousand years. It blanketed much of west-central India, known as the *Deccan Traps,* with layers of basalt hundreds of feet thick. The eruptions might have dealt a major blow to the climatic and ecological stability of the planet and possibly had a hand in the extinctions at the end of the period.

The Seychelles Bank was once part of India 65 million years ago, during a time when the subcontinent was drifting toward southern Asia (FIG. 7-3). The rift separated the Seychelles Bank from the mainland, leaving the islands behind as India continued to trek northward. Off the Seychelles Bank is what appears to be a 200-mile-wide impact crater that lies in the Amirante Basin, south of the Seychelles Bank about 300 miles northeast of Madagascar.

It is possible that a massive meteorite impact in the Amirante Basin triggered the great lava flows that created the Deccan Traps and the Seychelles Islands. Quartz grains, shocked by the high pressures generated by the impact, were found lying just beneath the immense lava flows, which suggests that they might be linked to the impact. Large asteroid impacts create so much disturbance in the Earth's thin outer crust they can induce massive volcanic eruptions.

Volcanologists argue that the shocked quartz and iridium found in the Cretaceous-Tertiary (K-T) boundary clay were derived from massive volcanic eruptions occurring over hundreds of thousands of years. Like large meteorite impacts, explosive volcanic eruptions, such as the 1980 eruption of Mount St. Helens in southwestern Washington (FIG. 7-4), can produce shocked quartz grains. However, the pressures generated by volcanic eruptions are not nearly as great as those generated by large asteroid impacts. Therefore, certain features, such as striations across shocked quartz grains, are not the same for volcanoes as they are for impacts.

Volcanoes whose magma source lies deep within the mantle, such as Mauna Loa on Hawaii (FIG. 7-5), also produce significant amounts of iridium, as well as osmium—another rare element found in the K-T boundary layer. Moreover, the microspherules found at the K-T contact that are believed to have been produced by an impact melt, could also have originated from volcanic eruptions. Therefore, the dinosaurs could have been killed without the help of an asteroid.

PLATE TECTONICS

Mass extinction events correlate reasonably well with cycles of terrestrial phenomena. The largest of these cycles is a 300-million-year cycle of convection currents in the Earth's mantle. *Convection* is the motion that occurs within a fluid medium as the result of a difference in temperature. Fluid rocks in the mantle receive heat from the core, ascend, dissipate their heat to the *lithosphere* (the outer rigid layer of the mantle and its overlying crust), cool, and descend to the core again to pick up more heat. The cycling of heat within the mantle is the main force behind *plate tectonics,* which is the movement of crustal plates on the

Earth's surface. These movements are responsible for all geologic activity occurring on the planet.

Plate tectonics have been operating since the early stages of the Earth and have played a prominent role in the history of life. Changes in the relative configuration of the continents and the oceans have influenced the environment, climate patterns, and the composition and distribution of species (FIG. 7-6). The continual changes in world ecology have had profound effects on the course of evolution and accordingly, on the diversity of living organisms.

During periods of rapid mantle convection, the supercontinents were broken up (FIG. 7-7). This separation led to the compression of ocean basins, rising sea

(PHOTO BY R.V. EMETAZ, COURTESY OF U.S. FOREST SERVICE)

Fig. 7-4. Area devastated by the explosive eruption of Mount St. Helens in spring 1980.

(COURTESY OF USGS)

Fig. 7-5. The Mauna Loa volcano on Hawaii is the largest shield volcano in the world.

levels and a transgressing the seas onto the land. Rapid mantle convection increases volcanism, which increases the carbon dioxide content of the atmosphere, resulting in a strong greenhouse effect with warm conditions worldwide. These episodes occurred from about 500 million to 350 million years ago and from about 200 million to 50 million years ago.

When the mantle convection was low, the landmasses assembled into a supercontinent. This connection led to a widening of the ocean basins, dropping global sea levels and a regressing the seas from the land. Moreover, the atmospheric carbon dioxide was reduced as a result of low levels of volcanism and the development of an "icehouse effect," which produces cooler temperatures worldwide. These conditions prevailed from about 700 million to 550 million years ago, from about 400 million to 250 million years ago, and during the latter part of the Cenozoic period.

Continental drift has had a profound effect on life on this planet since it first began. Continents and ocean basins are continuously being reshaped and rearranged by several crustal plates that are constantly in motion. When continents break up, they override ocean basins, make the seas less confined and thus raise global sea levels several hundred feet. Low-lying areas inland of the continents are inundated by the sea, dramatically increasing the shoreline and the shallow-water marine habitat area. The increased inhabitable area can thus support a larger number of species.

A great deal of mountain-building is associated with the movements of crustal plates. These mountains alter patterns of river drainage and climate, which in turn affect terrestrial habitats. Elevating land, where the air is thinner and colder, allows glaciers to grow, especially in the higher latitudes. Continental placement throughout the world can also interfere with ocean currents (FIG. 7-8), which affect global heat distribution.

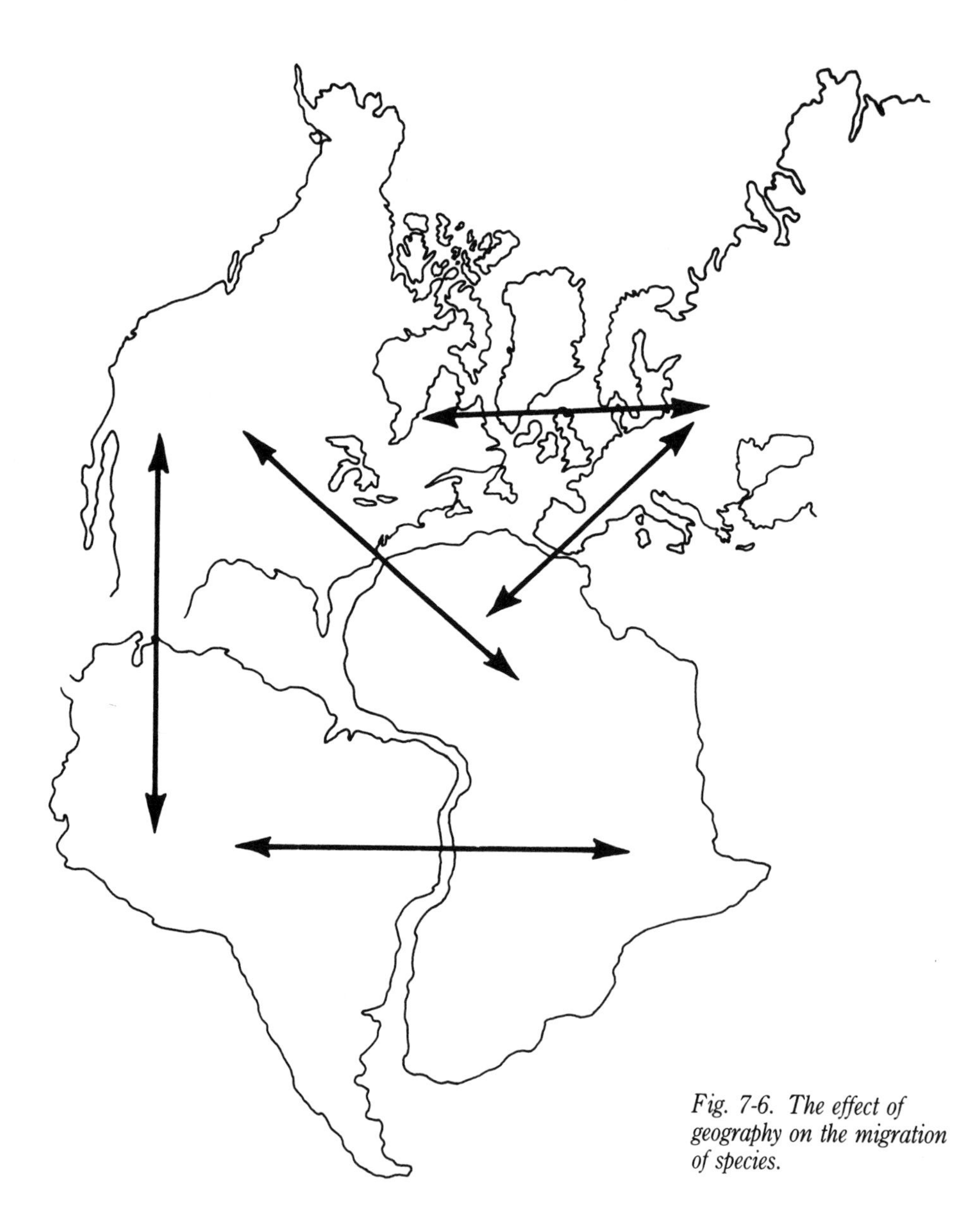

Fig. 7-6. The effect of geography on the migration of species.

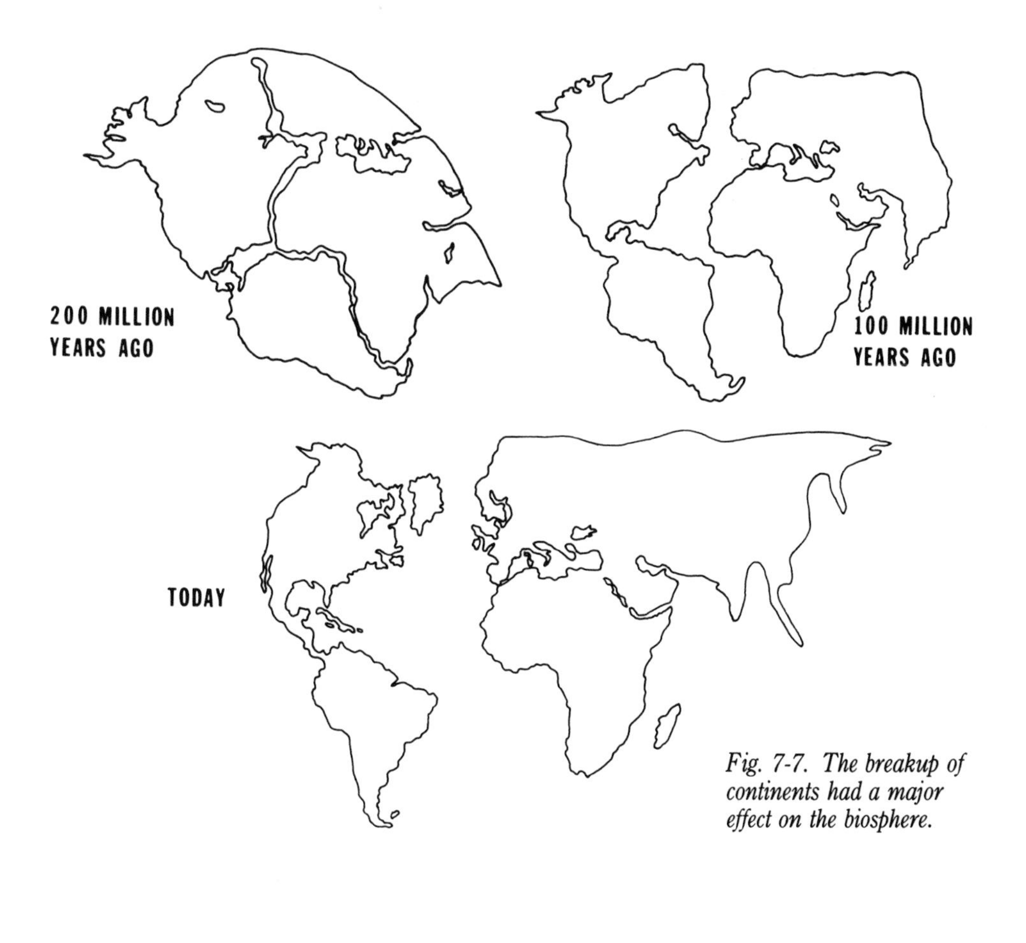

Fig. 7-7. The breakup of continents had a major effect on the biosphere.

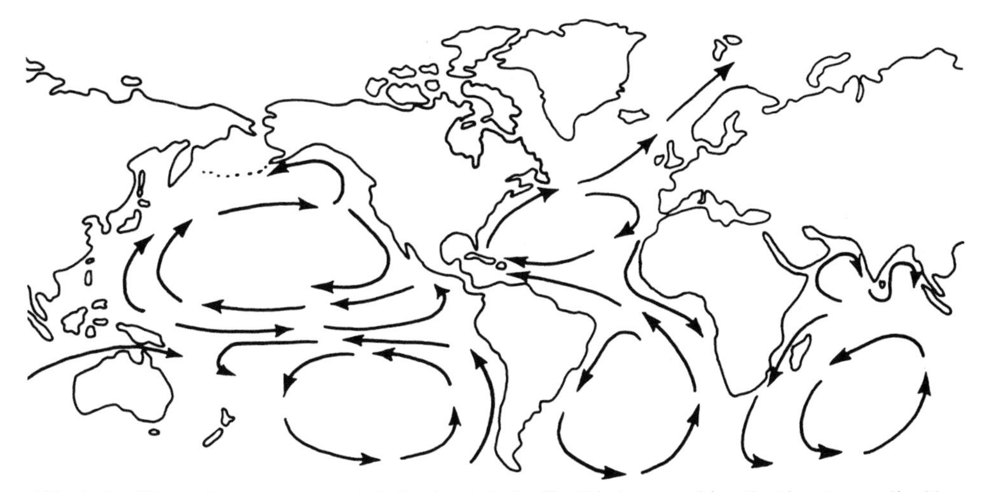

Fig. 7-8. The major ocean currents help circulate the Earth's heat, making the planet more livable.

When continents are assembled into a supercontinent, land no longer impedes free-flowing ocean currents, which distribute heat from the tropics to the poles and keep the temperatures of the planet more uniform. The ocean basins also widen, causing the sea level to drop considerably. A substantial and rapid fall in sea level could have both a direct and an indirect influence on the biological world. It would cause seasonal extremes of temperature on the continents to increase, thereby increasing environmental stress on terrestrial species.

A drop in sea level would also force the inland seas to retreat, producing a continuous, narrow continental margin around the supercontinent. This margin would reduce the shoreline, which would radically limit the marine habitat area. Moreover, unstable near-shore conditions would result in an unreliable food supply. Many species are unable to cope with the limited living space and food supply and die out in large numbers. Such an episode occurred at the end of the Permian period, when the vast majority of marine species became extinct.

GEOGRAPHICAL INFLUENCES

When all of the continents were welded into the supercontinent Pangaea (FIG. 7-9) near the end of the Paleozoic era around 250 million years ago, a great diversity of plant and animal life existed on the land and in the sea. With the large landmass located near the tropics, more of the Sun's heat was absorbed, which contributed to higher global temperatures. Oceans existing in the high latitudes are less reflective than land and absorb more heat, which further moderates the climate. Moreover, by not having land located in the polar regions to interfere with the movement of warm ocean currents, both poles remained ice-free around the year, with no large variation in temperature between the high latitudes and the tropics.

The formation of Pangaea marked a major turning point in life, during which time the reptiles emerged as the dominant species. It is believed that most of the Pangaean climate was equable and fairly warm throughout the year. However, much of the interior of Pangaea was a desert, whose temperatures fluctuated wildly from season to season, with scorching summers and freezing winters. This climate might have contributed to the widespread extinction of land-based species during the late Paleozoic. It also explains why the reptiles, which adapt readily to hot, dry climates, replaced the amphibians as the dominant land species.

During the breakup of Pangaea, beginning around 180 million years ago, the climate, particularly in the Cretaceous period, was extremely warm, and average global temperatures were 10 to 25 degrees Fahrenheit warmer than they are today. When the continents drifted toward the poles at the end of the Cretaceous, however, they disrupted the transport of poleward oceanic heat (TABLE 7-1) and replaced heat-retaining water with heat-losing land. As the cooling progressed, the land accumulated snow and ice, creating a greater reflective surface, which further lowered the global temperature and sea level.

Most marine species live, on continental shelves or in shallow-water portions near islands and subsurface rises, at depths generally less than 600 feet. Furthermore, extinctions tend to increase toward shore because shallow-water environ-

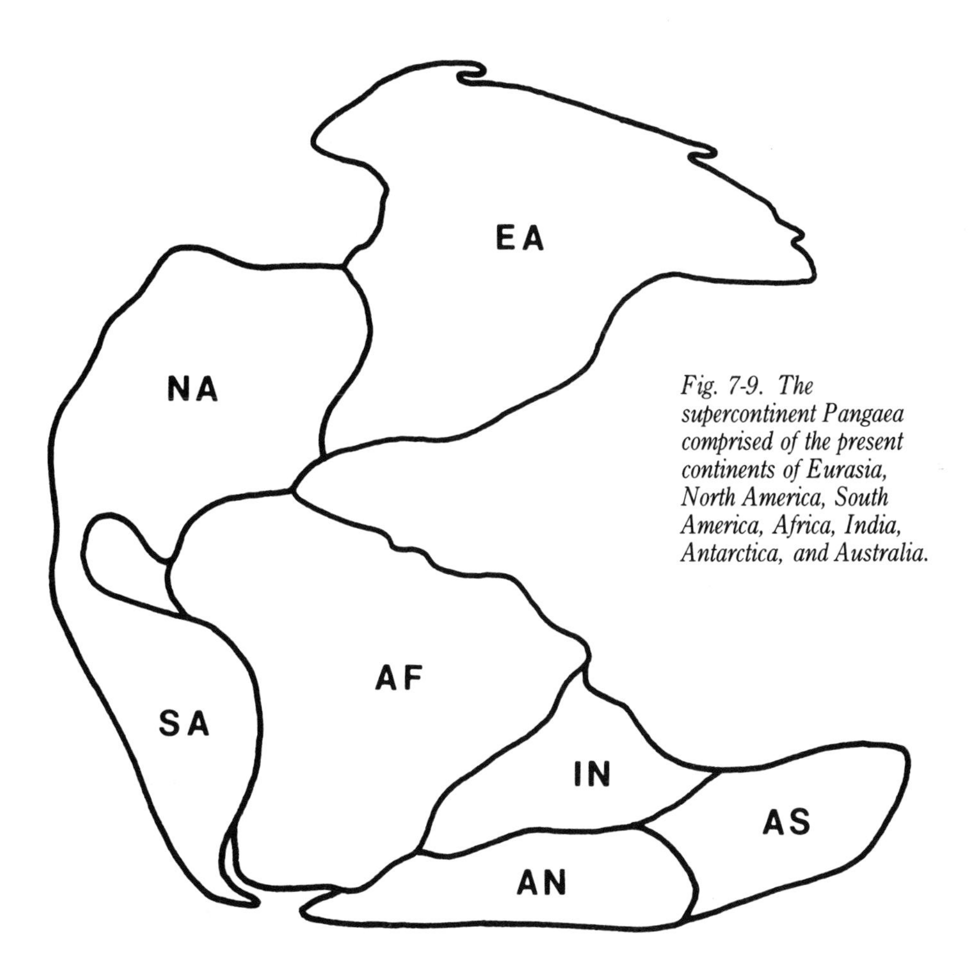

Fig. 7-9. The supercontinent Pangaea comprised of the present continents of Eurasia, North America, South America, Africa, India, Antarctica, and Australia.

TABLE 7-1. History of the Deep Circulation in the Ocean

AGE	EVENT
Less than 50 million years ago	The ocean could flow freely around the world at the equator. Rather uniform climate and warm ocean even near the poles. Deep water in the ocean is much warmer than it is today. Only alpine glaciers on Antarctica.

TABLE 7-1. *Continued.*

AGE	EVENT
35-40 million years ago	The equatorial seaway begins to close. There is a sharp cooling of the surface and of the deep water in the south. The Antarctic glaciers reach the sea with glacial debris in the sea. The seaway between Australia and Antarctica opens. Cooler bottom water flows north and flushes the ocean. The snow limit drops sharply.
25-35 million years ago	A stable situation exists with possible partial circulation around Antarctica. The equatorial circulation is interrupted between the Mediterranean Sea and the Far East.
25 million years ago	The Drake Passage between South America and Antarctica begins to open
15 million years ago	The Drake Passang is open; the circum-Antarctic current is formed. Major sea ice forms around Antarctica which is glaciated, making it the first major glaciation of the Modern Ice Age. The Antarctic bottom water forms. The snow limit rises.
3-5 million years ago	Artic glaciation begins.
2 million years ago	An Ice Age overwhelms the Northern Hemisphere.

ments fluctuate much more than environments further offshore. The richest shallow-water faunas are in the tropics, which contain large numbers of highly specialized species. Progressing to higher latitudes, the diversity gradually falls, until in the polar regions less than 10 percent as many species exist as in the tropics. Moreover, twice the species diversity exists in the Arctic sea, which is surrounded by continents, than is in the Antarctic sea, which surrounds a continent.

Species diversity mostly depends on the food supply, and as the seasons become more pronounced in the higher latitudes, the food production fluctuates. Diversity is also affected by seasonal changes, such as variations in surface and upwelling ocean currents that affect the nutrient supply, which in turn causes large fluctuations in productivity. Therefore, the greatest diversity among species is off the shores of small islands or small continents facing large oceans, where fluctuations in the nutrient supply are least affected by the seasonal effects of landmasses.

Diversity is also highly dependent on the shape of the continents, the width of shallow continental margins, the extent of inland seas, and the presence of coastal mountains, all of which are affected by continental motions. When the continents were assembled into Pangaea, a continuous shallow-water margin ran around it, with no major physical barriers to the dispersal of marine life. The seas were largely confined to the ocean basins and did not extend significantly over the continental shelves. Consequently, habitat area for shallow-water marine organisms was very limited, which accounted for the low species diversity. As a result, marine biotas were more widespread, but contained comparatively fewer species.

A similar circumstance might have occurred during the late Precambrian around 600 million years ago, when another supercontinent appears to have been in existence. During the Cambrian, it broke into four or five continents, which might have had a major affect on the explosion of new species during that time. When Pangaea broke up and the resulting continents migrated to their present positions, diversity again increased to unprecedented heights, providing our present-day world with a rich variety of species.

MAGNETIC REVERSALS

The Earth's magnetic field protects life against dangerous particle radiation from the Sun and from cosmic rays originating in outer space. The *solar wind*, composed of subatomic particles streaming outward from the Sun, stretches the magnetic field lines, creating the *magnetosphere* (FIG. 7-10). This layer in turn shields the Earth from deadly ionizing radiation. Without this protection, the bombardment of radiation could influence the composition of the upper atmosphere by generating higher levels of nitrogen oxides, which in turn would produce a haze to block out the Sun.

Geologic evidence taken from sequences of volcanic rock on the ocean floor, which record the polarity of the Earth's magnetic field when they cool and solidify, shows that the geomagnetic field has reversed itself many times in the past. After a long stable period lasting hundreds of thousands of years, the

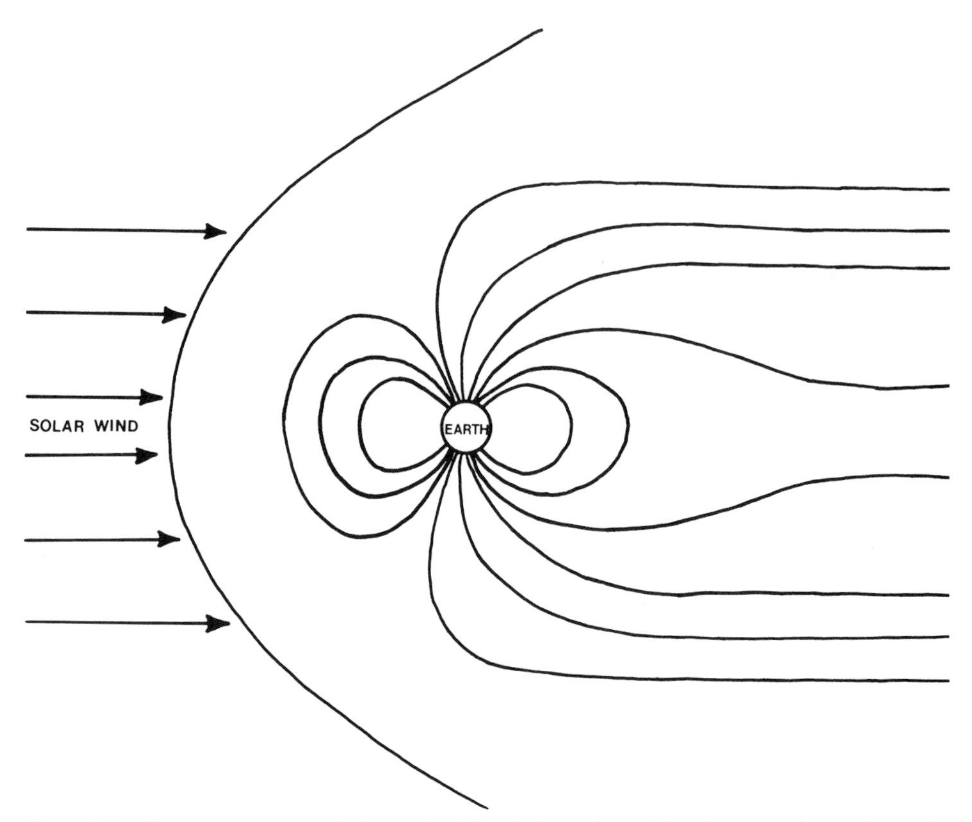

Fig. 7-10. The magnetosphere helps protect Earth from harmful solar particles and cosmic radiation.

strength of the magnetic field gradually decays over a short period of several thousand years. At some point, the field collapses all together, and a short time later it regenerates with the opposite polarity.

A comparison between magnetic reversals with variations in the climate has shown a striking agreement in many cases. Certain magnetic reversals also coincide with the extinction of species. Magnetic field reversals might be triggered by variations in the galactic magnetic field as the Solar System moves through the midplane of the galaxy. However, the galactic magnetic field appears to be too weak to influence the Earth's magnetic field, which is a million times stronger. Large meteorite impacts, very strong earthquakes, or intense volcanic activity have also been cited as causes for geomagnetic field reversals.

Magnetic field reversals have also been blamed for the ice ages, and a reversal occurring around 2 million years ago might have initiated the Pleistocene glaciation. Reversals in the magnetic field and excursions of the magnetic poles appear to correlate with periods of rapid cooling and the extinction of species. The Gothenburg geomagnetic excursion, which occurred about 13,500 years ago in the midst of a longer period of rapid global warming toward the end of the last ice age, resulted in plummeting temperatures and advancing glaciers for one thousand years. Apparently this excursion was caused by a weakened magnetic field.

Presently, the Earth's magnetic field is experiencing a slow, steady decrease in intensity. If this rate continues, it could collapse in the not too distant future. A second variation is a slow westerly drift of eddies in the field, amounting to one degree of longitude every five years. The drift suggests that the fluid in the outer metallic core (FIG. 7-11), which generates the geomagnetic field by the so-called dynamo effect, is moving at a rate of about a hundred yards a day.

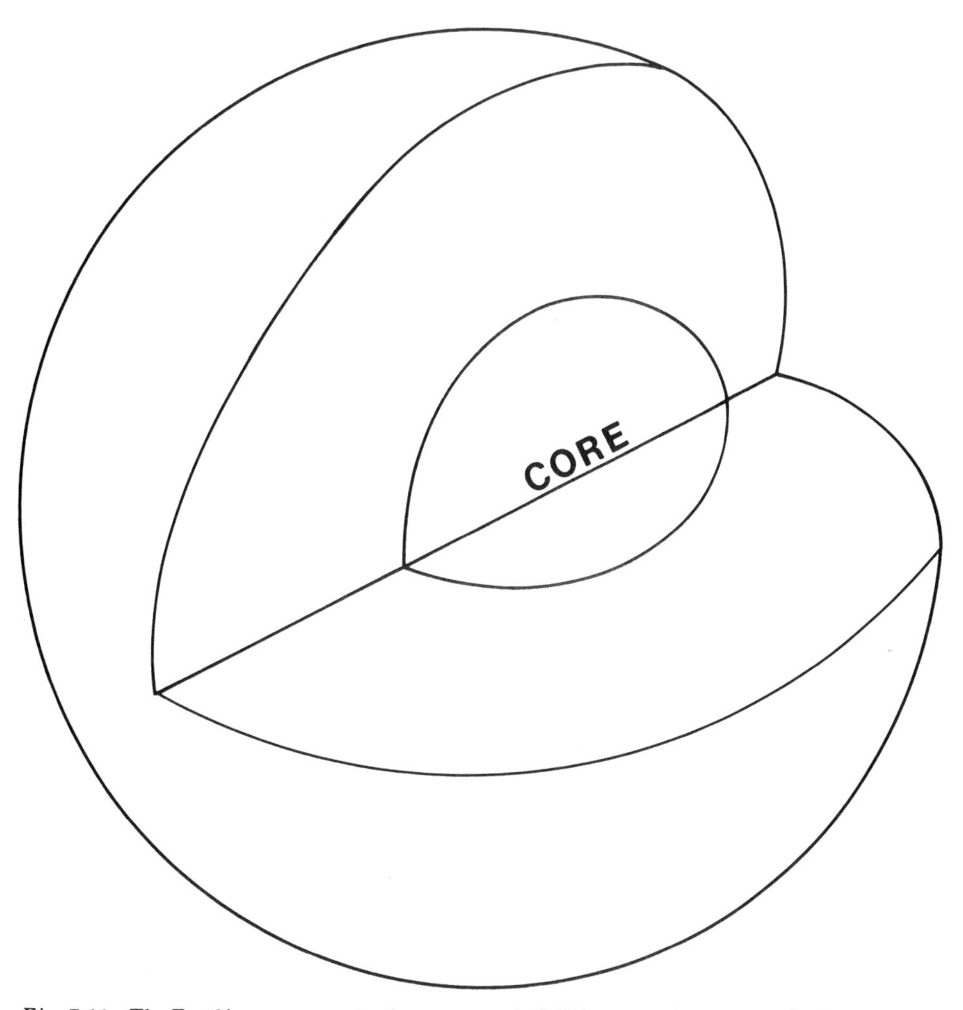

Fig. 7-11. The Earth's core generates the geomagnetic field by convection currents in the outer fluid layer.

The pattern of magnetic reversal is highly irregular and appears to result from a random process. The magnetic field reverses roughly two or three times every million years. Over the last 170 million years, it has reversed nearly 300 times. The last time the magnetic field reversed itself was about 700 thousand years ago, around the time when the Toba Volcano in Indonesia erupted—the greatest eruption of the past million years. The massive basalt floods of Long Valley, California occurred during this time. Also, a major meteorite landed in Australasia at this time. Ever since then, the Earth's climate has fluctuated between glacial and interglacial periods. Furthermore, the timing of these events suggests that the Earth is well overdue for another reversal.

Magnetic reversals have occurred throughout geologic time. Moreover, no single polarity has been dominant for long durations, except possibly during the Cretaceous period (between 135 and 65 million years ago). During this time an interval with no reversals appears to have existed for 35 million years. Also, toward the end of the Cretaceous, a long period of magnetic stability was interrupted by an abrupt reversal. During this period, the dinosaurs were declining.

Generally after several hundred thousand to over one million years of stability, the Earth's magnetic field strength suddenly drops over a period of less than two thousand years, followed by a delay of about 20,000 years. Then it abruptly collapses, reverses, and slowly builds back to its normal strength. Upwards of a thousand years might pass before the field regains its full magnetic intensity.

When the magnetic field collapses, it does not always reverse itself; in fact, 50 percent of the time it regenerates with the same polarity. It also appears that the Earth is experiencing a gradual decline in magnetic field intensity, which if it continues could lead to a reversal perhaps in another couple of thousand years—about when the next ice age is expected.

CLIMATIC COOLING

Possibly the most important factor to influence the diversity of species is climatic cooling. As the world's oceans cool, mobile species tend to migrate into the warmer regions of the tropics. Those species that are unable to move or those that become trapped in enclosed basins generally are the ones hardest hit. Only species that have previously adapted to cold conditions still thrive in today's oceans. Most of these are plant-eaters that tend to be generalized, rather then specialized feeders, consuming many types of vegetation.

Not all climatic cooling results in glaciation, which relies on several factors, such as the positioning of the continents, the tilt of the rotational axis, and the ellipticity of the Earth's orbit (FIG. 7-12). Nor did all extinctions follow sea-level lowering caused by the growing glaciers. In the Oligocene epoch, which began about 37 million years ago, seas that overrode the continents were drained when the ocean withdrew to one of its lowest levels in several hundred million years. Although the sea level remained depressed for five million years, little or no excess extinction of marine life occurred. Therefore, crowding conditions brought

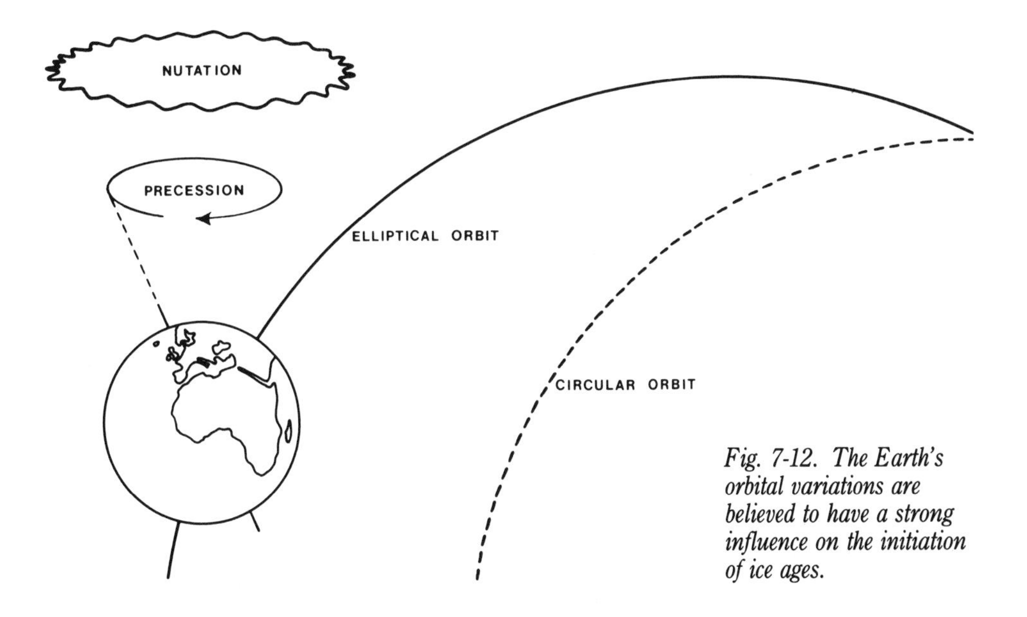

Fig. 7-12. The Earth's orbital variations are believed to have a strong influence on the initiation of ice ages.

on by lowering sea levels cannot be responsible for all extinction periods. Moreover, during many mass extinctions, the sea level was not much lower than it is today.

During the final stages of the Cretaceous period, when the seas were departing from the land and the level of the ocean began to drop, the temperatures in the broad tropical Tethys Sea began to fall. This droppage might explain why the Tethyan species that were the most temperature-sensitive suffered the most at the end of the period. Species that were amazingly successful in the warm waters of the Tethys were totally decimated when temperatures dropped. Afterwards, Marine species took on a more modern appearance as ocean bottom temperatures continued to plummet.

National Park Service

The "dinosaur family tree" portrays some of the world's many dinosaurs.

National Park Service

Allosaurus prowls a bone-littered riverbank.

National Park Service

The small plant eater Camptosaurus was an ancestor of many later dinosaurs.

National Park Service

The giant plant eater Camarasaurus.

National Park Service

Preparation of an Allosaurus fossil at the Dinosaur National Monument, Utah.

National Museums of Canada

Allosaurus in its natural habitat.

National Museums of Canada

A skeleton of a Hadrosaur, which is believed to have migrated in great herds.

Stegosaur on display at the Museum of Geology, South Dakota School of Mines at Rapid City.

8

The Ice Ages

EPISODES of extinction coincide with periods of glaciation, and the effect of global cooling on life is considerable. The living space of warmth-loving species is dramatically reduced to the narrowly confined tropics. Species also become trapped in confined waterways, unable to move to warmer seas. The accumulation of glacial ice in the polar regions lowers the sea level, and thereby reduces shallow water shelf areas. This limited habitat area in turn limits the number of species that can be supported. For this reason, major extinctions generally follow periods of climatic cooling. Moreover, a regular 26-million-year climatic fluctuation, together with changes in marine salinity and oxygenation, correlates with the extinction pattern over the last 180 million years.

Ocean temperature is by far the strongest controller of the geographic distribution of marine species, and climatic cooling is the primary culprit behind most of the extinctions in the ocean. Species unable to migrate or adapt to colder conditions are usually the ones hardest hit. Tropical species, especially, can only tolerate a narrow range of temperatures and have nowhere else to migrate to. Since lowered temperatures also slow the rate of chemical reactions, biological activity during a major glacial event would be expected to function at a lower energy state, which in turn could affect species diversity.

Not only have ice ages dramatically affected life on Earth for the last two billion years or so, but it is even possible that life itself has changed the climate sufficiently to initiate glaciation. Organisms possibly did this early in the Earth's history when they began substituting oxygen for carbon dioxide in the atmosphere, thereby weakening the greenhouse effect and lowering global temperatures. Life clearly affects the composition of the atmosphere and the oceans; without it, the Earth's climate would be totally out of control.

THE EARLY ICE AGES

The first glaciation recorded in the geologic record occurred during the early Proterozoic, two billion years ago (TABLE 8-1). This period was in transition, when atmospheric carbon dioxide was in the process of being replaced by oxygen generated by photosynthetic plants. The first microscopic plants developed

TABLE 8-1. Chronology of the Major Ice Ages

TIME IN YEARS	EVENT
2 billion	First major ice age.
700 million	The great Precambrian ice age.
230 million	The great Permian ice age.
230–65 million	Interval of warm and relatively uniform climate.
65 million	Climate deteriorates, poles become much colder.
30 million	First major glacial episode in Antarctica.
15 million	Second major glacial episode in Antarctica.
4 million	Ice covers the Arctic Ocean.
2 million	First glacial episode in Northern Hemisphere.
1 million	First major interglacial.
100,000	Most recent glacial episode.
20,000–18,000	Last glacial maximum.
15,000–10,000	Melting of ice sheets.
10,000–present	Present interglacial.

photosynthesis as early as 3.5 billion years ago. With this technique, they began to slowly replace the carbon dioxide in the ocean and in the atmosphere with oxygen.

The loss of carbon dioxide, which is an important greenhouse gas, resulted in substantial global cooling. This occurred even while the Sun was becoming progressively hotter. The oxygen produced by photosynthesis was removed by chemical processes and buried in the crust. This removal was fortunate, because the earliest organisms lacked defenses against the toxic effects of oxygen and would surely have perished in their own waste products.

About 2 billion years ago, these oxygen traps contained all the oxygen they could possibly hold, and the gas began building up in the ocean and in the atmosphere. Along with the generation of oxygen, simple plants removed carbon dioxide from the environment through photosynthesis. Furthermore, nonbiological chemical reactions stored carbon dioxide in carbonate rock on the bottom of the ocean. These carbon dioxide repositories dramatically reduced the greenhouse effect and lowered global temperatures.

Meanwhile, plate tectonics became a powerful force in shaping the surface of the planet. The stored carbon in the carbonaceous sediments, along with the oceanic crust on which they were deposited, were thrust deep inside the Earth (FIG. 8-1). The growing continents also stored great repositories of carbon in thick deposits of carbonate rock, suck as limestone. The elimination of carbon dioxide in this manner substantially cooled the Earth. Moreover, the early Proterozoic continental landmass, which by this time occupied about 25 percent of the Earth's surface, might have been located near one of the poles, where ice sheets grow easily.

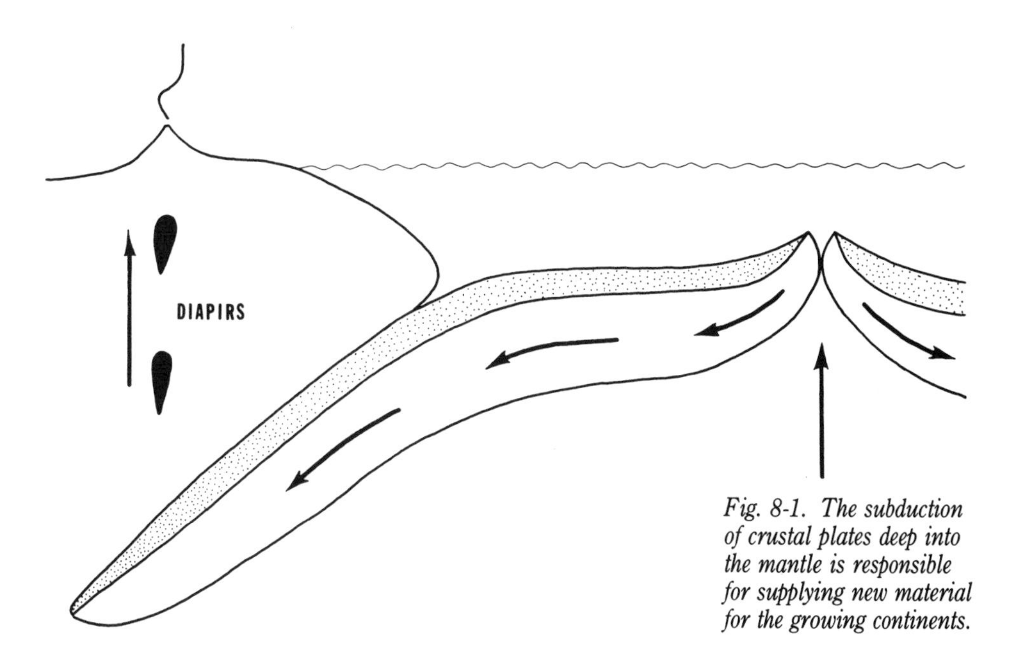

Fig. 8-1. The subduction of crustal plates deep into the mantle is responsible for supplying new material for the growing continents.

The worst period of glaciation in Earth history occurred during the late Precambrian (around 670 million years ago), when nearly half the land surface at that time was covered with ice (FIG. 8-2). The global climate was so cold that ice sheets and permafrost existed even at equatorial latitudes. During this time, no plants grew on the barren landscape and only simple one-celled plants and animals lived in the sea.

Fig. 8-2. Location of late Precambrian glacial deposits throughout the world.

Toward the end of the Precambrian, a supercontinent located near the equator broke apart and one of the continents might have wandered into a polar region and became blanketed with thick ice. The ice age dealt a terrible blow to life in the ocean, and it is believed that many simple organisms vanished during this time. The extinction decimated the ocean's population of acritarchs, a single-celled phytoplankton that was the first organism to evolve cells with nuclei.

By the time the glaciation ended and the ice sheets retreated, life began to proliferate with an intensity never shown before or since. Three times as many phyla (organisms that share the same general body plan) existed then as do today. As a result, many unique and bizarre creatures dominated the fossil record of that time (FIG. 8-3).

THE ANCIENT ICE AGES

The movement of the continents across the face of the Earth is thought to be responsible for another period of glaciation during the late Ordovician (around 440 million years ago). The trilobites and 100 other families marine animals were struck down by the ice age. During this time, North Africa hovered directly over the South Pole and ice sheets flowed across the continent in all directions.

Another marine crisis occurred near the end of the Devonian period (about 370 million years ago), when tropical species were again hit hard. No major extinction event occurred during the widespread Carboniferous glaciation (around 330 million years ago), however. The relatively low extinction rates probably reflect that the ecosystem had not yet fully recovered from the late Devonian extinctions. Nevertheless, one very consistent pattern of mass extinctions throughout geologic history is that even though each event typically affects different suites of organisms, tropical biotas containing the largest number of species were nearly always the hardest hit.

The Carboniferous glaciation along with the Permo-Carboniferous glaciation (around 290 million years ago), might have been influenced by a reduction of atmospheric carbon dioxide to about 25 percent of its present value. Geologic evidence taken from deep-sea cores drilled into the ocean crust indicates that carbon dioxide variations preceded changes in the extent of the more recent ice ages. It is therefore assumed that the earlier periods of glaciation were similarly affected.

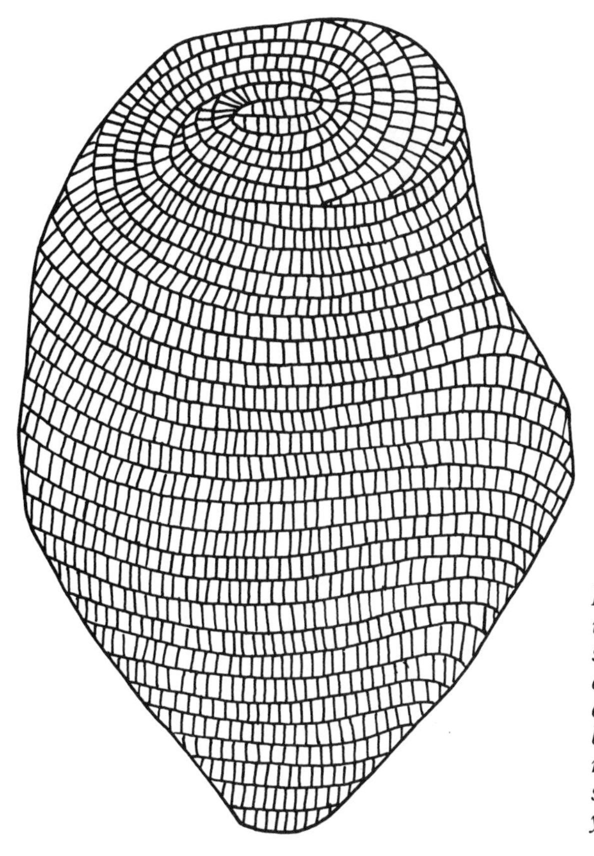

Fig. 8-3. Helicoplacus, which was an experimental species whose parts were assembled different from every other living creature, became extinct about 510 million years ago after surviving for 20 million years.

The variations in carbon dioxide levels might not be the sole cause of glaciation, however. When combined with other processes, such as variations in the Earth's orbital motions or a drop in solar radiation, the fluctuations in carbon dioxide levels could spell the difference between whether the Earth is glacier-free or coated with ice.

The great forests that spread across the land during the latter part of the Paleozoic era provided a substantial carbon dioxide repository. Extensive coal deposits, including those in the eastern and central United States (FIG. 8-4), were created from these forests. Plants invaded the land and extended to all parts of the world, beginning about 450 million years ago. Lush forests that grew during the Carboniferous period used their woody tissues to store large quantities of carbon dioxide, which were later deposited in the crust. Buried under layers of sediment, the vegetative matter was compacted and converted into thick seams of coal.

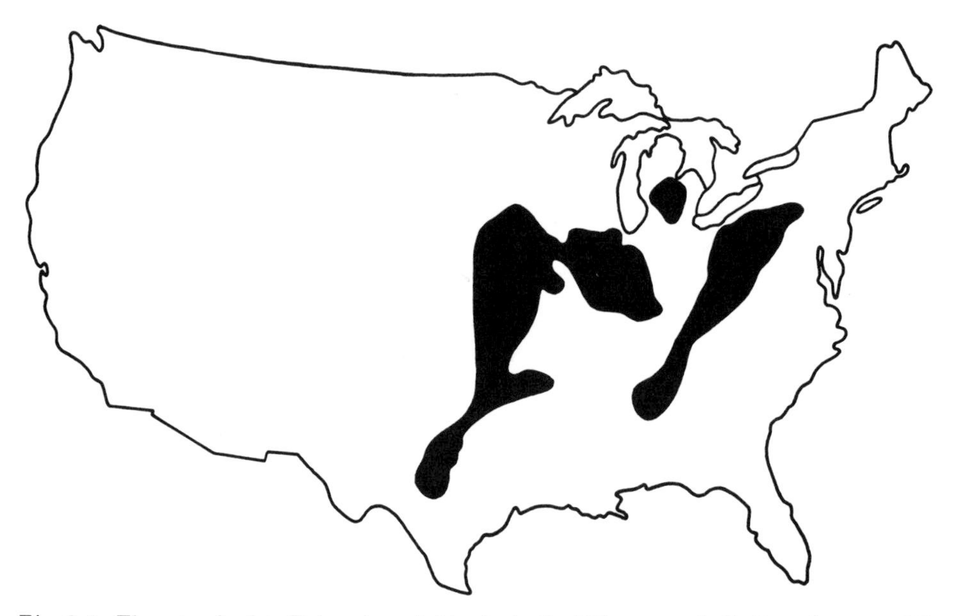

Fig. 8-4. The extensive late Paleozoic coal deposits in the U.S. are an indication of once-prolific forest growth.

This was also a time of extensive mountain building, and great chunks of crust were raised to higher elevations, where glaciers were nurtured in the cold, thin air (FIG. 8-5). Glaciers might have formed and persisted on continents even at low latitudes as long as a high elevation was maintained. With higher altitudes, the temperatures decrease and the precipitation increases.

During the latter part of the Permian period (around 290 million years ago), Gondwana drifted into the south Polar regions, where glacial centers expanded across the continents (FIG. 8-6). Ice sheets covered large portions of east central South America, South Africa, India, Australia, and Antarctica. Land existing near

(PHOTO BY A. POST, COURTESY OF USGS)

Fig. 8-5. White Chuck Glacier at the Glacier Peak Wilderness area, Snohomish County, Washington.

the poles often causes extended periods of glaciation. This is because land located at higher latitudes usually has a high *albedo* (reflective properties), and a low heat capacity, which encourages the accumulation of ice.

When Gondwana and Laurasia merged into the supercontinent Pangaea around 250 million years ago, the continental collisions crumpled the crust and pushed huge masses of rocks into several mountain chains around the world (FIG. 8-7). In addition to these folded mountain belts, volcanoes were also highly active. Unusually long periods of volcanic activity might have culminated in thick ash clouds that blocked out the Sun and lowered surface temperatures.

As the plates collide, the crumpling of the crust causes the continents to rise higher. Meanwhile, the ocean basins drop lower. All known episodes of glaciation occurred when sea levels should have been low; however, not all mass extinctions were associated with lowered sea levels. The changes in the shapes of the ocean basins greatly affected the course of ocean currents, which in turn had a pronounced effect on the climate. The continental margins became less extensive and narrower, confining marine habitats to near-shore areas. Such an occurrence might have had a major influence on the great extinction at the end of the

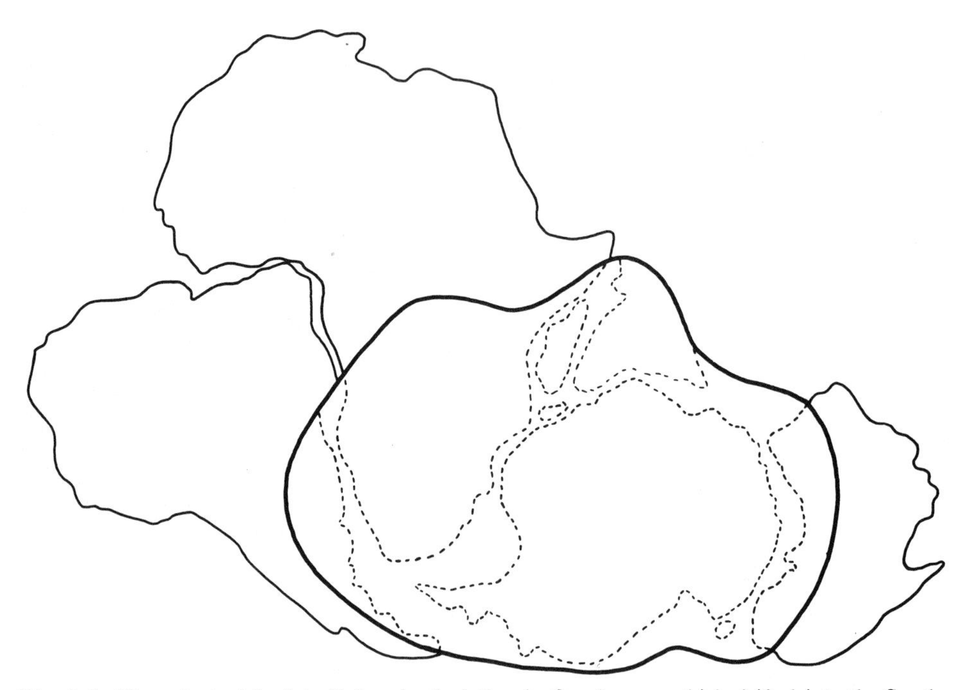

Fig. 8-6. The extent of the late Paleozoic glaciation in Gondwana, which drifted into the South Polar region during this time.

Fig. 8-7. The major mountain ranges of the world resulted mostly from continental collisions.

Paleozoic era. During this time, land that was once covered by great coal swamps completely dried when the climate grew colder.

CRETACEOUS ICE

During the Cretaceous period, the warmest period of the Earth's history, plants and animals were particularly abundant and widespread. Volcanoes were especially active during this time and injected massive amounts of carbon dioxide into the atmosphere. As a result of the greenhouse effect, the heated up planet might have been responsible for the prodigious plant growth and the giantism of the dinosaurs.

The average global surface temperature, which presently is about 60 degrees Fahrenheit, was 20 to 25 degrees warmer during the Cretaceous. The deep ocean waters, which now hover near freezing, were 30 degrees warmer. The polar regions were also much warmer than they are today; there is no evidence that any permanent glaciers existed during this time. The temperature difference between the poles and the tropics was only 40 degrees—about half of what it is today.

No major ice sheets existed during the Cretaceous. However, large, out-of-place boulders strewn across the desert in central Australia is evidence for the existence of small amounts of glacial ice. During this time, the interior of Australia was filled with a large inland sea. Sediments settling on the floor of the basin were

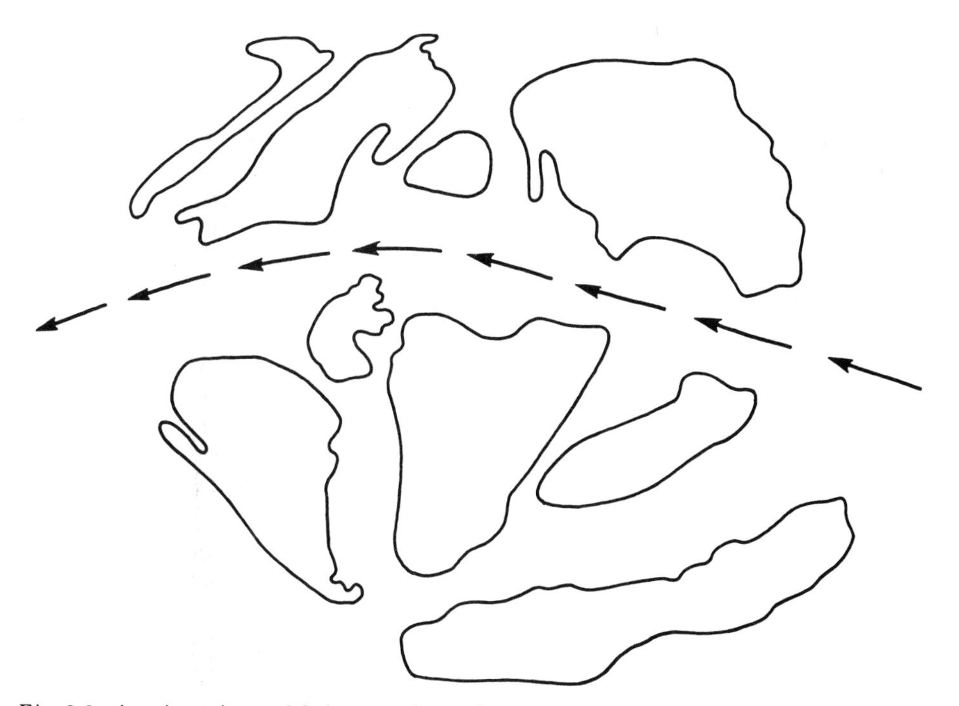

Fig. 8-8. A unique circumglobal current in the Tethys Seas helped provide the warm temperatures of the Cretaceous.

lithified into sandstone and shale, which later became exposed when the sea departed at the end of the period. Lying in the middle of the sedimentary deposits are curious-looking boulders of exotic rock, called *drop stones.* These boulders came from a great distance and yet measure as much as 10 feet across.

The appearance of these strange boulders in "the middle of nowhere" suggests that they were rafted out to sea on slabs of sea ice. When the ice melted, the huge rocks simply dropped to the seafloor, where their impacts disturbed the underlying sediment layers. Apparently during the middle Cretaceous, Australia, which was still attached to Antarctica, wandered near the Antarctic Circle. During the cold winters, portions of the interior coastline froze into pack ice. Rivers of broken ice then flowed into the inland sea, carrying the embedded boulders that were dropped over 60 miles from shore.

When the Cretaceous ended, the seas regressed from the land as the sea level lowered and the climate generally grew colder. The last stage of the Cretaceous, called the *Maestrichtian,* was the coldest interval of the period. No clear evidence exists to prove that significant glaciation occurred at this time. However, most warmth-loving species, especially those living in the Tethys Sea (FIG. 8-8), disappeared when the Cretaceous ended. The extinctions apparently occurred gradually—over a period of one to two million years. Moreover, the species that were already in decline (including the dinosaurs), might have been dealt a final death blow when a large asteroid impacted on the Earth. This impact might have clogged the skies with dust and further lowered global temperatures.

THE RECENT ICE AGES

Over the last 100 million years, extinction events have coincided to some extent with three major steps in the evolution of the Cenozoic climate. The first was the onset of the mid-latitude Northern Hemisphere glaciation about 2.4 to 3.0 million years ago (FIG. 8-9), the second was a major expansion of ice on Antarctica between 10 and 14 million years ago, and the third was a major cooling between about 31 and 40 million years ago. Also, a forth extinction event coincides with a major environmental change that occurred about 90 million years ago.

Near the end of the Eocene epoch (about 37 million years ago), global temperatures dropped significantly. Antarctica, which had separated from Australia and wandered over the South Pole, acquired a thick blanket of ice. Glaciers also grew for the first time in the highest ramparts of the Rocky Mountains (FIG. 8-10), which were uplifted during the *Larimide orogeny* (mountain building episode) between 80 million and 40 million years ago. Moreover, the wide Tibetan Plateau, which soars above three miles, was uplifted during this time as India continued to shove against southern Asia. Mass extinctions in the ocean, eliminated many types of plankton and those on the land forced the disappearance of the arachaic mammals.

A permanent ice cap did not develop over the North Pole until about four million years ago when Greenland acquired its first major ice sheet (FIG. 8-11). During this time, the Panama Isthmus was uplifted, impeding the flow of ocean currents from the Atlantic into the Pacific. This event stranded many species—

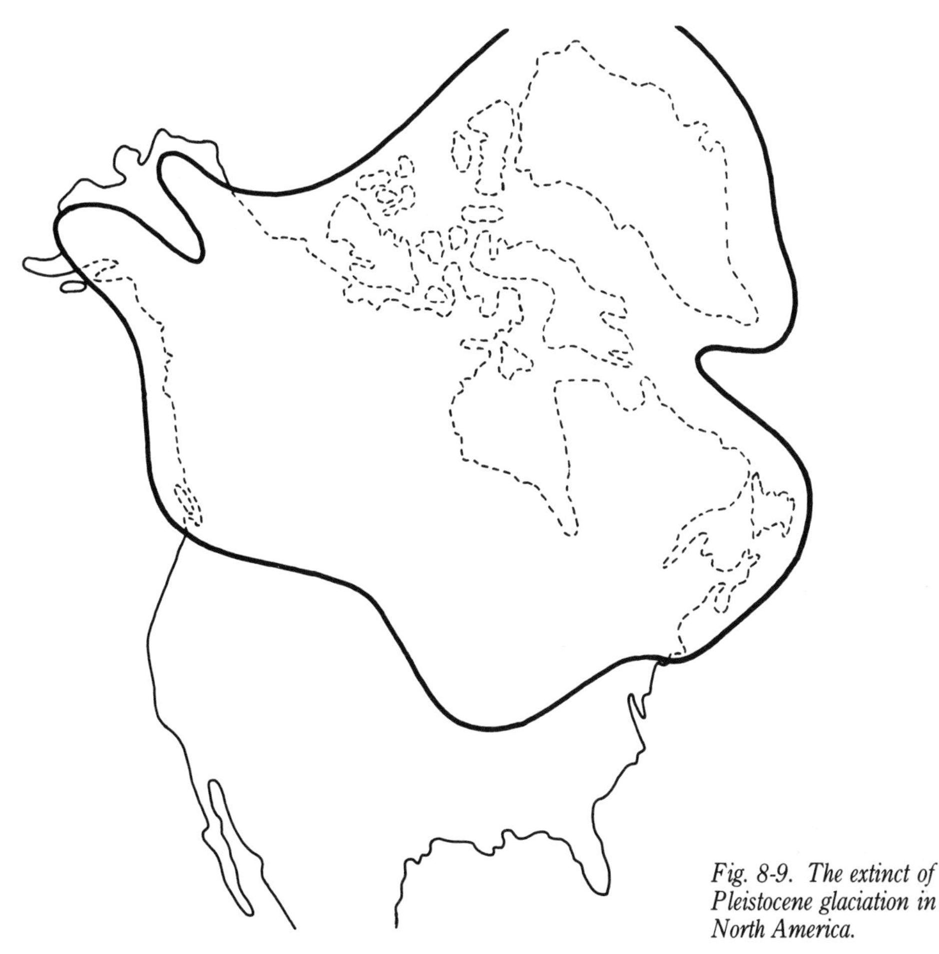

Fig. 8-9. The extinct of Pleistocene glaciation in North America.

they were unable to escape the cold waters of the Atlantic Ocean and it subsequently led to their extinction. Of the marine fossil faunas that are four million years old, more than 50 percent of the species are extinct. Faunas of the western Atlantic and the Caribbean were the hardest hit, losing about 70 percent of the mollusk species. In the Mediterranean and the North Sea, about 30 percent of these species were wiped out.

Populations of one-celled algae, called *diatoms* (FIG. 8-12), whose shells are composed of silica, sharply decreased in the Antarctic surface waters 2.4 million years ago. During this time, sea ice extended its reach northward and shaded the algae below. Without sunlight for photosynthesis, the diatoms simply vanished. The timing of the disappearance of the diatoms is generally accepted as the beginning of the Pleistocene glaciation. It was a most unusual period in that no major mass extinctions occurred, as had happened during many of the earlier glacial episodes.

During the Pleistocene, a progression of ice ages spanned across the northern continents. Each ice age was followed by a short interglacial period, similar to the one we are experiencing today. During warm interglacial spells, species invade all latitudes. However, as glaciers advance across the continents and ocean temperatures drop, species crowd into warmer regions, where habitats and food supplies are limited.

The last ice age began about 100 thousand years ago, intensified about 75 thousand years ago, peaked about 18 thousand years ago, and retreated about 10 thousand years ago. At the height of this ice age, five percent of the planet's water was locked in glacial ice. This ice lowered the sea level by as much as 400 feet and expanded the land area by as much as eight percent. So, the shallow-water shelf areas were reduced by about the same amount.

Coral, which lives only in warm surface waters, fluctuates in height in response to changing sea levels. The shallow sea caused the coral reefs to erode

(PHOTO BY W. CROSS, COURTESY OF USGS)

Fig. 8-10. Snow-covered peaks of the Rocky Mountains near Telluride, Colorado.

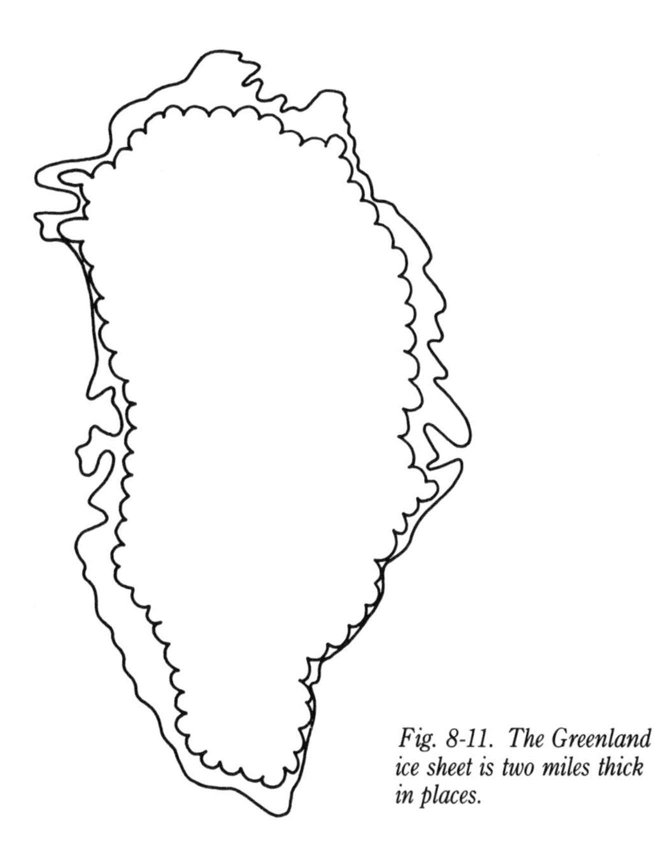

Fig. 8-11. The Greenland ice sheet is two miles thick in places.

to the new sea level. When the sea rose as a result of the melting glaciers, new coral began to grow on top of the old. When the corals were dated using radiometric dating techniques, they provided a reliable chronology for the ice ages. A comparison between oxygen isotopes in the fossil coral also yielded the mean temperature, which generally was about 10 degrees Fahrenheit lower than it is today.

The cold weather and advancing ice forced species to migrate to warmer latitudes. Ahead of the slowly advancing ice sheets, which covered perhaps only a few hundred feet per year, lush deciduous forests gave way to evergreen forests. These in turn yielded to grasslands, which completely disappeared to become barren tundra and rugged periglacial regions that existed at the margins of the ice sheets (FIG. 8-13).

Precipitation rates fell dramatically as lowered temperatures caused less water to be evaporated from the oceans. Because very little melting occurred in the cooler summers, only minor amounts of snowfall were necessary to sustain the ice sheets. The lower precipitation levels also increased the spread of deserts across many parts of the world. Desert winds were much more blustery than they are today, producing dust storms of gigantic proportions. So much dust was

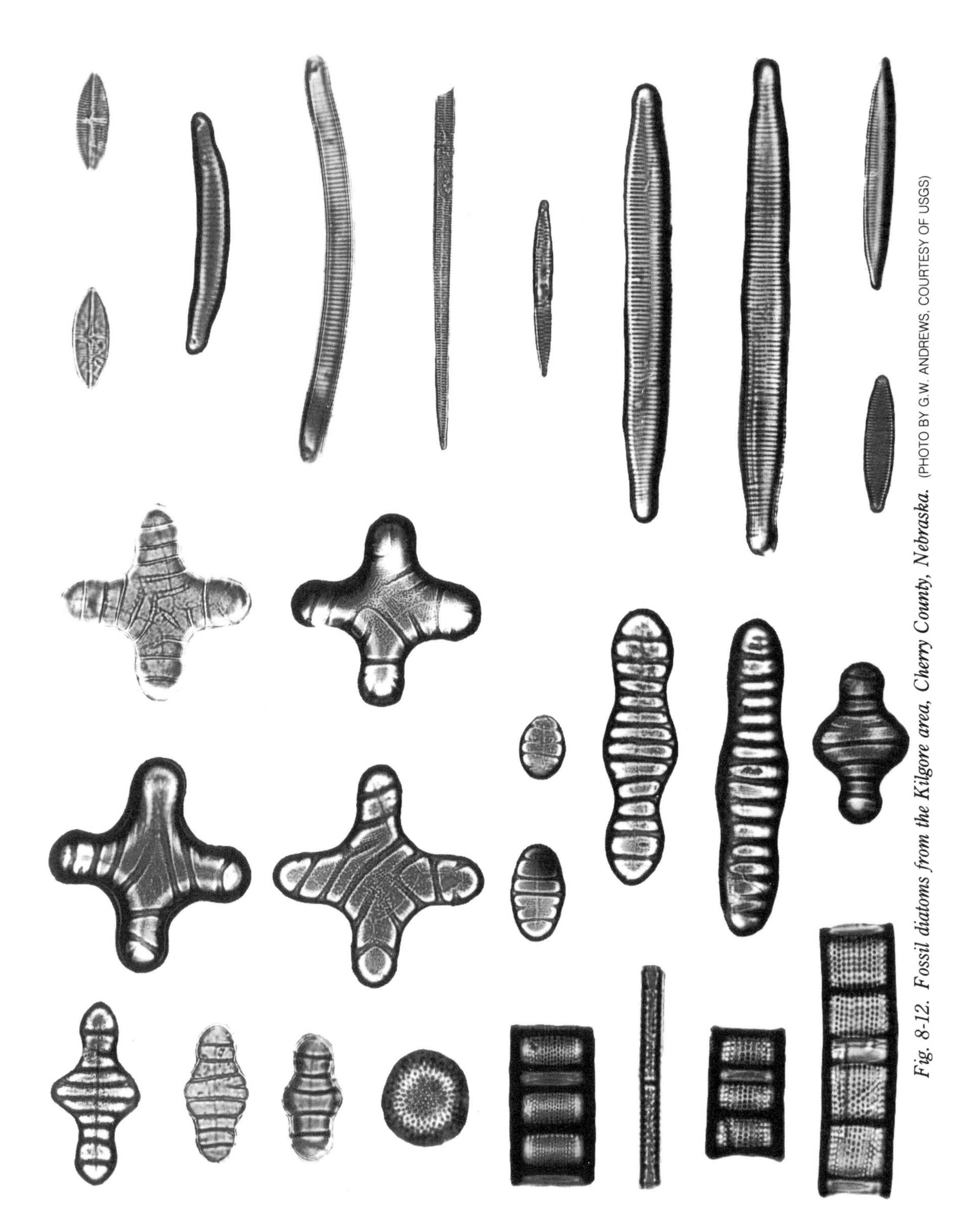

Fig. 8-12. Fossil diatoms from the Kilgore area, Cherry County, Nebraska. (PHOTO BY G.W. ANDREWS, COURTESY OF USGS)

(COURTESY OF NATIONAL PARK SERVICE)

Fig. 8-13. Periglacial terrain near Sperry Glacier, Glacier National Park, Montana.

suspended in the atmosphere that it significantly blocked out sunlight, thus shading the Earth and keeping it cool.

THE PRESENT INTERGLACIAL

Perhaps one of the most dramatic climate changes in geologic history occurred during the present interglacial, the Holocene epoch. After 90,000 years of gradual accumulation of snow and ice (up to two miles thick in the higher latitudes of North America and Eurasia), the glaciers melted away in only a few thousand years, retreating upwards of 600 yards annually. Following the receding ice sheets, plants and animals began to return to the north.

About 30 percent of the ice melted between 16 and 13 thousand years ago, when the average global temperature increased about 10 degrees Fahrenheit to nearly what it is today. Around 13 thousand years ago on the border between Idaho and Montana, a gigantic ice dam that held back a huge lake hundreds of miles wide and up to two thousand feet deep, suddenly burst, and the waters gushed toward the Pacific Ocean. Along the way, they carved out some of Earth's strangest landscapes, known as the *Scablands* (FIG. 8-14).

(PHOTO BY F.O. JONES, COURTESY OF USGS)

Fig. 8-14. The Scablands of the northern end of the upper Grand Coulee were formed when floodwaters from a break in a gigantic ice dam tore through the landscape toward the Pacific Ocean.

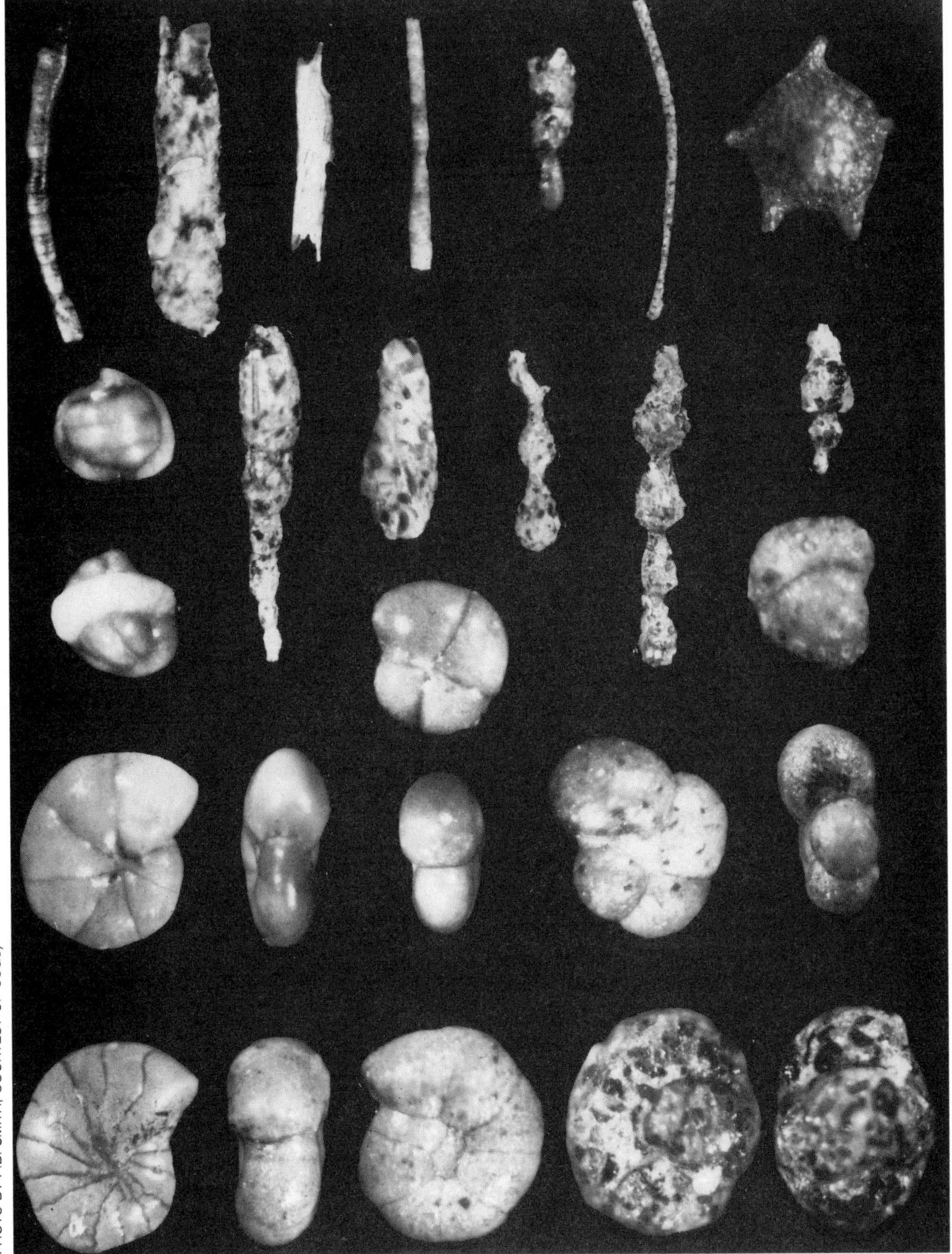
(PHOTO BY P.B. SMITH, COURTESY OF USGS)

Fig. 8-15. Foraminifera of the North Pacific Ocean.

When the North American ice sheet began to retreat, its meltwater flowed down the Mississippi River, into the Gulf of Mexico. After the ice sheet retreated beyond the Great lakes, however, the meltwater took a different route down the St. Lawrence River, and the cold waters entered the North Atlantic Ocean. The rapid melting of the glaciers culminated in the extinction of microscopic organisms, called *foraminifera* (FIG. 8-15). Their demise was brought about when a torrent of meltwater and icebergs spilled into the North Atlantic. This mixture formed a cold freshwater lid over the ocean that significantly changed the salinity of the seawater. The cold waters also blocked poleward flowing warm currents from the tropics, and caused the land temperatures to fall to near ice-age levels.

As the glaciers started to retreat, the cool, stable climate of the ice age gave way to the warmer, more turbulent climate of the present interglacial. The rapid environmental switch from glacial to interglacial shrunk the forests in favor of grasslands. This pattern might have disrupted the food chains of several large mammals, causing them to become extinct.

Almost simultaneously, 35 classes of mammals and 10 classes of birds became extinct in North America. The extinctions occurred between 12 and 10 thousand years ago and peaked around 11 thousand years ago. The vast majority of the mammals affected were large plant-eaters that weighted over 100 pounds, with some weighing as much as a ton or more. Unlike earlier episodes of extinction, this one did not significantly affect small mammals, amphibians, reptiles, and marine invertebrates.

Archaeologists discovered that from 11.5 to 11 thousand years ago, many parts of North America were occupied by ice age peoples, whose spear points were found among the remains of giant mammals, including mammoths, mastodons, tapirs, native horses, and camels. These people migrated to North America from Asia, when lowered sea levels exposed a land bridge across the Bering Strait. They moved through the ice-free corridor east of the Canadian Rockies and entered a land populated with upwards of 100 million large mammals.

The changing climate, resulting from the transition out of the ice age, culminated with rising sea levels and falling water levels over much of the North American continent. The large mammals probably congregated at the few remaining water holes, where they became vulnerable to hunting pressures by humans. With such plentiful prey and exposure to few, it any, new diseases, human population growth exploded and people spread over the continents, one step behind the retreating glaciers.

9

Greenhouse Warming

THE world is in danger of overheating as a result of excess greenhouse gases in the atmosphere. This added greenhouse warming is, in large measure, caused by the trapping of incoming solar radiation by man-made pollutants—especially carbon dioxide and methane. However, the climate change theory is still not complete enough to provide all the answers concerning the importance of the greenhouse effect. More research is needed on the atmosphere, along with its interactions with the ocean. Important information about the Earth still needs to be collected, and the best way to do this is with advanced satellite technology. When this information is analyzed, a decade or more might be required to sort it.

Perhaps, if the current upward temperature trend continues well into the next decade, scientists will be more certain that it is tied to the greenhouse effect. However, it is uncertain whether the human race, or the rest of the living world for that matter, has enough time to take evasive action. If we wait too long to install corrective measures, much more drastic steps might be required to counter global warming in the future. Moreover, lead times for building greenhouse-combating projects, such as pollution-free solar power plants, might require a decade or more. The thermal inertia of the ocean might delay the onset of greenhouse warming for several decades, by which time it could hit with full fury.

THE PUZZLING CLIMATE

The decade of the 1980s witnessed the six hottest years of the century, which even surpassed the Dust Bowl years of the 1930s (FIG. 9-1). Extreme and often

(PHOTO BY A. ROTHSTEIN, COURTESY OF USDA-SOIL CONSERVATION SERVICE)

Fig. 9-1. The 1930s Dust Bowl in Oklahoma.

record-breaking weather events have occurred worldwide, including heat waves in American cities and Central European capitals, floods in Africa that interrupted nearly two decades of drought, and almost continuous rain and cold in the middle of summer at other places.

These events might well be symptoms of a global climate change caused by the chemical pollution of the atmosphere. However, climatic variability is such that the strange weather could simply be a reflection of natural variation. As yet, no clear sign of climatic change can be positively blamed on the greenhouse effect. Also, unknown moderating factors might cancel part of the greenhouse effect. It still remains a mystery where all the carbon dioxide produced by industrial activities is going. It appears that only 50 percent of the carbon dioxide generated by the combustion of fossil fuels and the destruction of the forests is accumulating in the atmosphere and in the ocean.

Marine single-celled plants, called *nannoplankton*, produce a gaseous sulfur that might help counter human-induced global warming by partially regulating the Earth's temperature. The sulfur gas emissions could increase the concentration of cloud-forming particles in the atmosphere, which could make clouds reflect more sunlight. This would in turn lower global temperatures. Other types of plankton might be encouraged to grow more vigorously by supplying the oceans with nutrients, thus allowing these tiny marine plants to absorb more carbon dioxide. Volcanic eruptions, decreasing solar activities, and decreasing stratospheric concentrations of ozone could also help cool the Earth.

The climate has always been changing, even without man's contribution. During the last warm interglacial period, which began about 125 thousand years ago, the atmospheric carbon dioxide content was greater than it is today, the climate was much warmer, and sea levels were 20 feet higher as a result of the melting ice caps. The climate 18 thousand years ago, at the height of the last ice age when glaciers covered about 30 percent of the northern landmass, was significantly colder than it is today. The climate of the last 200 years has been warmer than the previous 200 years, a period known as the *Little Ice Age.* And the climate of the last 50 years has been even warmer than that of the previous 50 years, when alpine glaciers were at their southernmost extent.

In the past, climate changes were slow enough for the biological world to adapt. However, today's climate changes are much too abrupt, which could cause plants and animals to become extinct. Global warming would be hardest on plants because they are directly affected by changes in temperature and rainfall. Forests, especially game preserves, would become isolated from their normal climate regimes, which would continue to move northward. Human intervention on an unprecedented scale might be required to preserve plant and animal species that are threatened by global warming, especially if it occurs too rapidly.

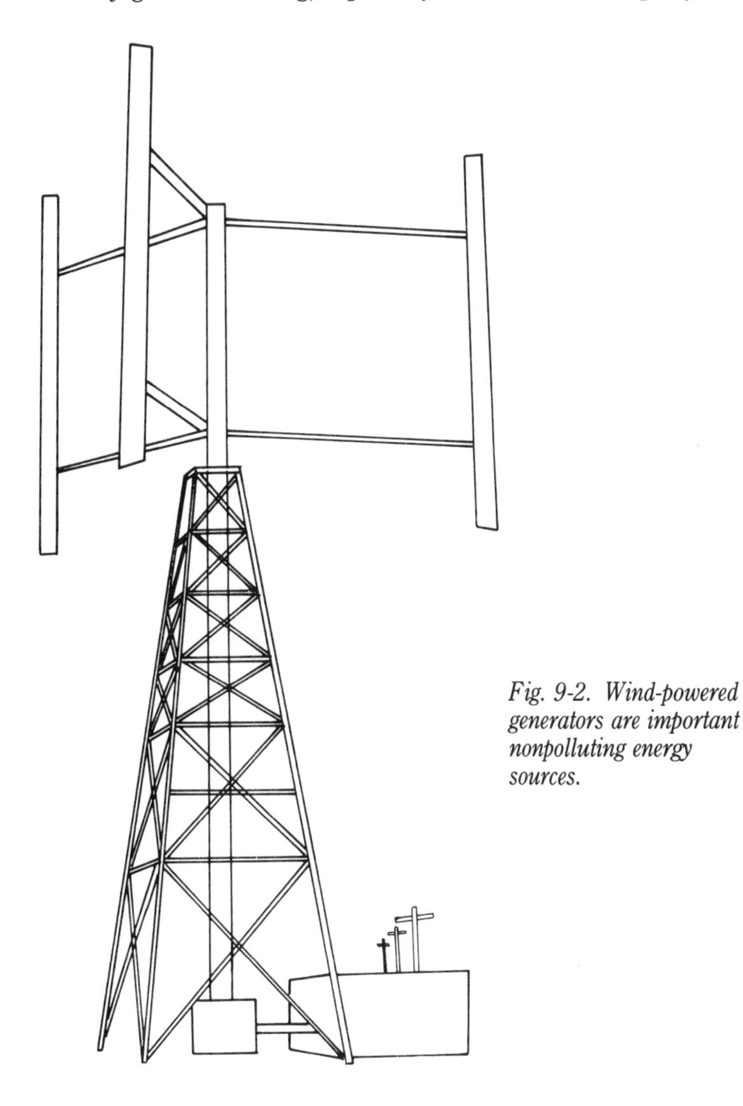

Fig. 9-2. Wind-powered generators are important nonpolluting energy sources.

Humans too, might find themselves threatened by climatic change. One response strategy for combating climatic change is adaptation, such as moving to a cooler climate or building coastal defenses against rising sea levels. Another response strategy is limitation, which involves limiting or reducing the emissions of greenhouse gases. A prudent response to climatic change is to utilize both these measures.

Conservation measures also could help curtail the effects of global warming. These steps would result in large part from the consequences of improved energy efficiency along with the development of nonpolluting substitute energy sources (FIG. 9-2). However, conservation can only attack a portion of the carbon dioxide problem. It will not solve the problem.

CARBON DIOXIDE

Sometime between 2020 and 2070, if present trends continue, the concentration of carbon dioxide in the atmosphere could double, and the global mean surface temperature could increase by about five degrees Fahrenheit and perhaps by as much as ten degrees in some areas. The carbon dioxide content of the atmosphere also fluctuates between seasons (FIG. 9-3), rising to a peak in late winter and falling to a minimum at the end of summer. This is because plants draw carbon dioxide from the atmosphere during the growing season.

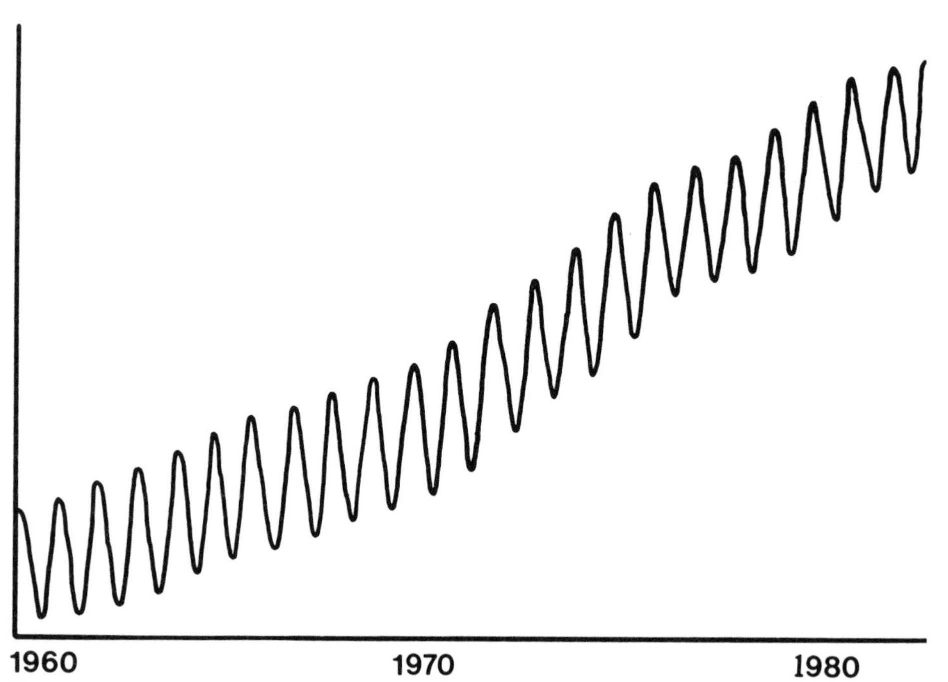

Fig. 9-3. The carbon dioxide content of the atmosphere fluctuates with the seasons and increases with time.

The forests of the world have a pronounced influence on the carbon dioxide content of the atmosphere. Much of the seasonal variation in the atmospheric concentration of carbon dioxide can be correlated with a rapid rise in photosynthesis during the summer. Forests conduct more photosynthesis worldwide than any other form of terrestrial vegetation. The stores of carbon in the forests are large enough to substantially affect the carbon dioxide content of the atmosphere.

Human activities are mostly responsible for the long-term increase in the atmospheric carbon dioxide content. If trends continue, the climate will generally become warmer over the next several decades. Although the mechanisms are complicated and not yet fully understood, the consequences of a steady rise in atmospheric carbon dioxide would probably be catastrophic if other moderating factors do not come into play. One such remedial action is the absorption of excess carbon dioxide and heat by the oceans. However, the transfer of atmospheric carbon dioxide into the oceans is slow. Moreover, the seas can only process 50 percent of the excess carbon dioxide generated by humans.

The carbon dioxide content of the atmosphere has increased by about 25 percent since 1860. This increase is mostly the result of an accelerated release of carbon dioxide by the combustion of fossil fuels. The present consumption of fossil fuels yields on average about 1.1 tons of atmospheric carbon for each of the world's 5.3 billion people yearly. Americans, who alone consume 20 percent of the world's natural resources, release nearly six tons of carbon per person per year. This amounts to 1.2 billion tons, or about 25 percent of the world's total.

The atmosphere presently holds about 700 billion tons of carbon. Therefore, we are increasing the amount of atmospheric carbon by about one percent annually. Some of the carbon is removed from the atmosphere by biospheric processes, so the average annual increase of atmospheric carbon dioxide by man's activity is cut in half, amounting to about three million tons.

The biota on the surface of the Earth and humus in the soil can hold 40 times more carbon than the atmosphere can. The harvest of forests, the extension of agriculture, and the destruction of wetlands speed the decay of humus and release vast quantities of carbon dioxide into the atmosphere. Deforestation by itself accounts for 15 to 30 percent of the global carbon dioxide emissions. Also, agricultural lands do not store as much carbon as the forests they replace. However, agricultural lands do release large amounts of carbon dioxide whenever the land is cultivated. The tilled land exposes humus to the atmosphere, where it becomes oxidized. Furthermore, young trees planted to replace those harvested for timber do not store as much carbon as older trees. For this reason, the world's ancient forests must be protected from destruction by timber interests.

The oceans contain the largest store of carbon dioxide, as much as 60 times more than the atmosphere. The carbon dioxide enters the ocean by the agitation of the surface water, and the concentration of carbon dioxide in the topmost 250 feet equals the amount in the entire atmosphere. Microorganisms living in this mixed layer of the ocean use carbon dioxide in the form of bicarbonate to make calcium carbonate for their skeletons and shells.

When the animals die, their skeletal parts settle on the shallow bottom, where they contribute to the formation of carbonate rock, such as limestone. If the calcium carbonate falls to greater depths, however, it is dissolved in the cold deep waters of the abyssal. This region of the ocean, by virtue of its great volume, holds the vast majority of free carbon dioxide. The upwelling of carbon dioxide-rich waters from great depths return the carbon dioxide to the atmosphere. For this reason, carbon dioxide concentrations are much higher around the equator, where the upwelling zones are, than at other latitudes.

Carbon dioxide enters the ocean from the atmosphere very slowly and at nearly a constant rate. Furthermore, this rate is only about 50 percent of the carbon dioxide that is generated by the combustion of fossil fuels. Therefore, without man's contribution, the atmosphere and the ocean would be in equilibrium—with the amount of carbon dioxide being absorbed by seawater equal to that supplied to the atmosphere. As a result of increasing concentrations of carbon dioxide, humans are short-circuiting the *carbon cycle*, the flow of carbon through the biosphere (FIG. 9-4).

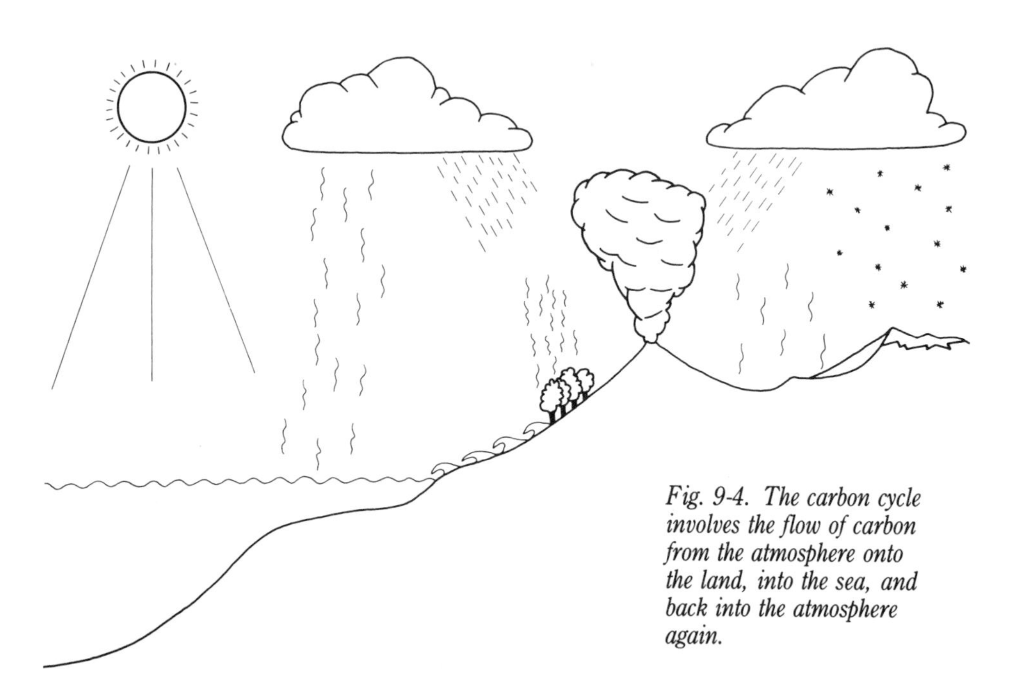

Fig. 9-4. The carbon cycle involves the flow of carbon from the atmosphere onto the land, into the sea, and back into the atmosphere again.

To increase surface temperature by doubling the amount of atmospheric carbon dioxide could be detrimental to worldwide precipitation. Areas between 20 and 50 degrees north latitude and 10 to 30 degrees south latitude could experience a marked decrease in precipitation, and encourage the spread of deserts. Presently the United States has more than 500 million acres of arid and semiarid land, which is about 20 percent of its total land area. Even larger desert areas exist in Africa, Australia, and South America (TABLE 9-1). Changes in precipitation patterns would have profound effects on the distribution of the water resources, especially in areas where they are desperately needed for human survival.

TABLE 9-1. Major Deserts of the World

DESERT	LOCATION	TYPE	AREA SQUARE MILES × 1000
Sahara	North Africa	Tropical	3500
Australian	Western/interior	Tropical	1300
Arabian	Arabian Peninsula	Tropical	1000
Turkestan	S. Central U.S.S.R.	Continental	750
North America	S.W. U.S./N. Mexico	Continental	500
Patagonian	Argentina	Continental	260
Thar	India/Pakastan	Tropical	230
Kalahari	S.W. Africa	Littoral	220
Gobi	Mongolia/China	Continental	200
Takla Makan	Sinkiang, China	Continental	200
Iranian	Iran/Afganistan	Tropical	150
Atacama	Peru/Chile	Littoral	140

Generally, however, rainfall is expected to increase globally. But not every region will receive an additional supply of rain, and some regions could actually become drier as evaporation rates increase as a result of the warmer climate. With diminished rainfall, higher temperatures, and increased evaporation, the flow of rivers could decline by 50 percent or more, and some rivers could dry out entirely. Major groundwater supplies could also be severely reduced, causing irrigation wells to go dry. Other areas could be swamped with an overabundance of precipitation, resulting in extensive flooding (FIG. 9-5).

RISING SEA LEVELS

The present rate of sea level rise is perhaps 10 times greater than it was 40 years ago. In most temperate and tropical regions of the world, the sea level is now rising about 0.25 inch per year. Most of the increase appears to come from melting glaciers in West Antarctica and Greenland, where there also appears to be a greater number of icebergs, which calve off glaciers entering the sea. The icebergs also appear to be getting larger. Alpine glaciers, which contain substantial quantities of ice, appear to be melting as well. Furthermore, most of the melting appears to be caused by global warming.

(PHOTO BY W. HOFFMAN, COURTESY OF USGS)

Fig. 9-5. Extensive flooding on the Feather River, Sutter County, California.

The increased temperature also produces a thermal expansion of the ocean, which increases its overall volume. Over this century, thermal expansion has raised the level of the sea by as much as two inches. This could result in an additional rise in global sea levels, which would alter the shapes of the continents and sink low-lying atolls and barrier islands. For every foot of sea level rise, 100 to 1000 feet of shoreline would disappear (FIG. 9-6). This is especially distressing for millions of people who live on low-lying fertile deltas and depend on these areas for their livelihood.

In some areas, the level of the sea has risen by as much as three feet per century. Louisiana is losing about six thousand acres of land each year to the encroaching sea. The beaches along North Carolina are retreating at a rate of 4 to 5 feet per year. The higher sea levels are caused in part, by the sinking of the land as a result of the increasing weight of seawater pressing down on the continental shelf. In regions like Scandinavia, the sea level has actually dropped by as much as three feet per century as a result of an increased buoyancy of the land since the glaciers departed after the last ice age.

The first possible signs that rising global temperatures have started to warm the ocean were revealed by satellite measurements of the extent of the polar sea ice, which shrank by as much as six percent during the 1970s and 1980s. Sea ice forms a frozen band around Antarctica that covers most of the Arctic Ocean during the winter season in each hemisphere (FIG. 9-7). If global warming melts the polar sea ice, the number of microscopic organisms would be reduced, and marine animals that feed on them would suffer as well. Less sea ice would also affect seals, which breed on the ice, and polar bears, which hunt and travel on the ice.

The world will feel the adverse effects of rising sea levels while the ice caps melt as a result of the rising sea temperatures. If the present melting continues, the sea could rise as much as six feet by the middle of the next century. Large

(PHOTO BY J. BISTER, COURTESY OF USDA-SOIL CONSERVATION SERVICE)

Fig. 9-6. Old stumps and roots exposed by shore erosion at Dewey Beach, Delaware, indicate that this area was once the tree zone.

(PHOTO BY E. MORETH, COURTESY OF U.S. NAVY)

Fig. 9-7. The U.S. Coast Guard icebreaker Polar State plows through sea ice near McMurdo Sound, Antarctica.

tracts of coastal land would disappear, along with low-lying islands, where many exotic species live. Delicate wetlands (FIG. 9-8), where many species of marine life nurture their young, would be reclaimed by the ocean. Vulnerable coastal cities would have to move inland or else build protective walls against the rising sea, on which a larger number of dangerous hurricanes spawn, as a result of the increase in global temperatures.

DROUGHT

The carbon dioxide that has been accumulating in the atmosphere over this century is expected to cause global warming by the greenhouse effect. Like in a greenhouse, escaping radiant heat is trapped by the atmosphere and reradiated back to the ground. The increase in global temperatures could dramatically alter the climate and shift precipitation patterns throughout the world. This alteration could bring unusually wet conditions to some areas and droughts to others. Moreover, both the frequency and the severity of droughts is likely to increase as a result greenhouse warming.

The African droughts of the 1980s, which left one million or more people dead or dying from famine, were the worst in this century. Part of the problem was caused by denuding the land of its vegetation, which altered its reflective properties. Another influence might have been the increase in atmospheric carbon dioxide by the combustion of fossil fuels, the destruction of forests and wetlands, and the extension of agriculture. Furthermore, the loss of vegetation reduces the amount of carbon dioxide that is taken from the atmosphere by green plants.

The central portions of continents that normally experience occasional drought could become permanently dry wastelands. The soils in almost all of Europe, Asia, and North America will become drier, requiring as much as 50

(PHOTO BY M.W. WILLIAMS, COURTESY OF NATIONAL PARK SERVICE)

Fig. 9-8. Mosquite Lagoon at Cape Canaveral National Seashore.

percent more irrigation (FIG. 9-9). Expected rises in temperatures, increased evaporation, and changes in rainfall patterns will severely limit the export of excess food to developing nations during times of famine.

(PHOTO BY G. ALEXANDER, COURTESY OF USDA-SOIL CONSERVATION SERVICE)

Fig. 9-9. An irrigated cornfield in eastern Nebraska.

Since the total heat budget of the Earth (FIG. 9-10) does not change significantly from year to year, areas that experience drought have their counterparts in areas that become unusually wet. During the decade of the 1980s, the United States experienced a series of bad droughts. Australia also had its most severe drought in over 100 years. An equally intense drought caused food shortages in southern Africa and affected West Africa and the Sahel region that borders the Sahara Desert as well. Meanwhile, the worst floods of the century struck South America in Ecuador, northern Peru, and large areas of Brazil, Paraguay, and Argentina.

Those regions that lie 30 degrees on either side of the equator can expect dramatic shifts in precipitation patterns as the world continues to heat up. The seasonal winds of the monsoons, which bring life-sustaining rainfall to about 50 percent of the people of the world, affect the continents of Asia, Africa, and Australia. However, perturbations in the climate, brought on by global warming, can lead to years of drought or flood, placing many people in great peril.

CLIMATE CHANGE

Greenhouse warming has a substantial effect on the climate (FIG. 9-11). Without greenhouse gases in the atmosphere, mean global temperatures would drop by about 60 degrees Fahrenheit and the planet would become a solid block

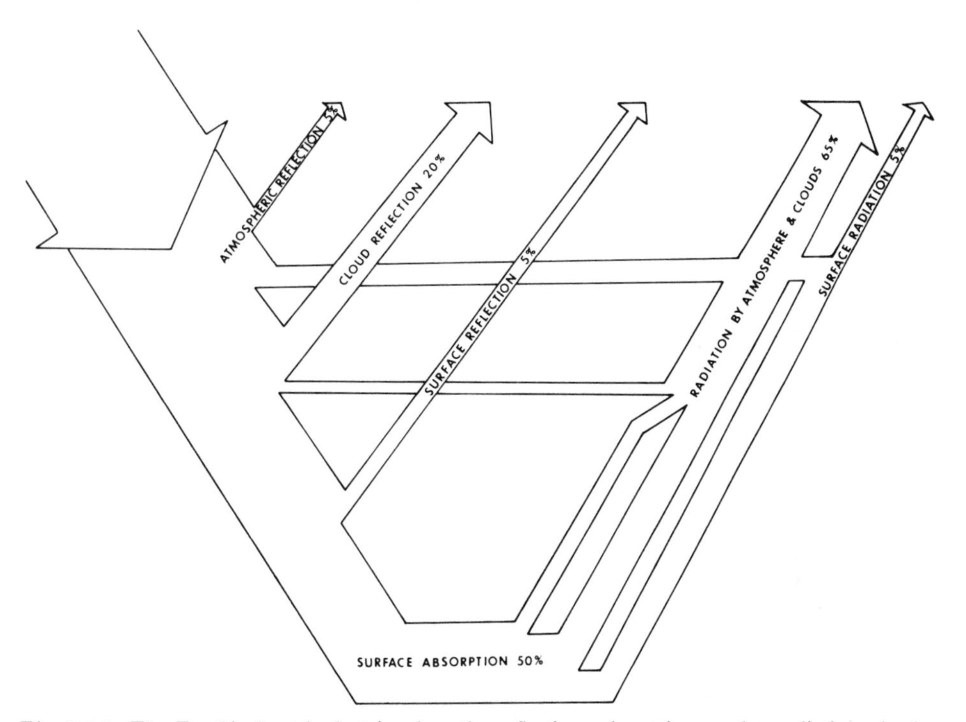

Fig. 9-10. The Earth's heat budget involves the reflection, absorption, and reradiation of solar insolation, all of which must escape to space in order to balance the temperature of the planet.

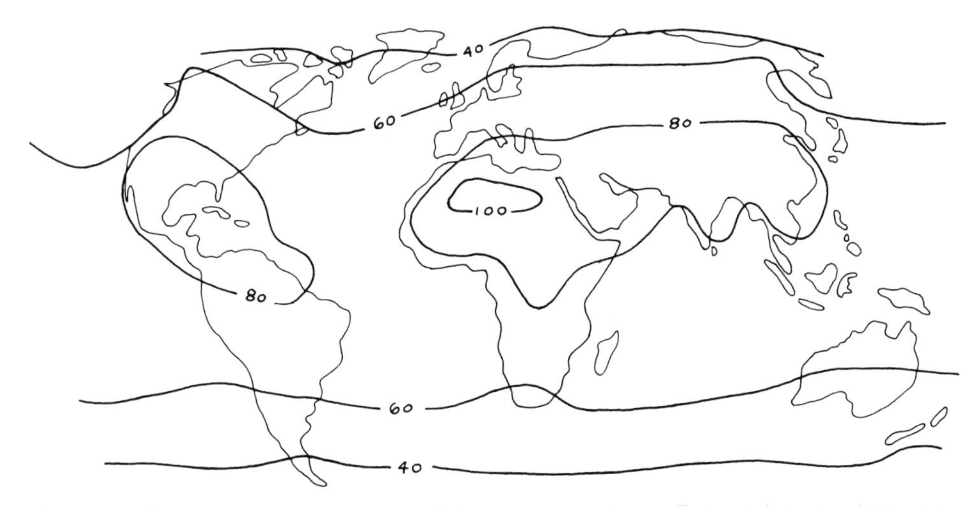

Fig. 9-11. Global mean temperatures could increase 5 to 10 degrees Fahrenheit by the middle of the next century if the present warming trend continues. Shown are today's mean temperatures in July.

of ice. Scientists have known the mechanics of the greenhouse effect for some time. However, few of them publically support the theory of global warming and most prefer a wait-and-see attitude.

If trends continue as they are, possibly by the middle of the next century, the Earth could be hotter than it has been in the past million years as a result of high atmospheric carbon dioxide levels and increases in other artificial greenhouse gases. Global warming will be felt the most at the higher latitudes of the Northern Hemisphere, where the largest temperature increases will occur during winter. Evaporation rates will increase, changing circulation patterns and dramatically affecting the weather. Once-productive croplands could lose topsoil and become man-made deserts, and many rain forests will be turned into deserts, as trees are cut down by the millions. From 50 to 90 percent of all species live in the tropical forests, which are vanishing at an atrocious rate of 40 acres per minute.

The present warming trend, amounting to a rise of over one degree Fahrenheit during this century, is remarkable for its unprecedented speed. The rate of global warming is approximately 40 times faster than it was at the end of the last ice age, when glaciers two miles thick began to melt. During the ending moments of that ice age, between 14 and 10 thousand years ago, the Earth warmed by as much as 5 to 10 degrees Fahrenheit, which is comparable to the temperature increase predicted for the greenhouse effect. The major difference, however, is that the last warming occurred over several thousand years, whereas the present warming trend will be compressed into less than a century.

If greenhouse warming continues on its present course, by the end of the next century, global temperatures could be as warm as they were 100 million years ago during the height of the Cretaceous, the hottest period in geologic history when the dinosaurs ruled the world. During this time, however, the continents were confined mostly to regions around the equator, so the climates of both periods would not be identical.

The higher temperatures will cause some areas, particularly those in the Northern Hemisphere, to dry out, creating a large potential for massive forest fires. During the relatively warm, dry 15th and 16th centuries, major forest fires in North America occurred roughly once every nine years. Over the next three centuries, during the cooling period of the Little Ice Age, massive forest fires were less frequent and less intense, occurring only about once every 14 years. This information has dire implications for us today. For if greenhouse warming continues, major forest fires might become more frequent and intense, greatly eliminating forests and their wildlife habitats.

As the result of global warming, which could continue for centuries, forests in the Northern Hemisphere would be forced to move further northward, while other wildlife habitats including those in the arctic tundra would disappear entirely (FIG. 9-12). The high northern latitudes are expected to warm to greater degree than regions further south. One horrifying possibility for the Arctic tundra is that the rising temperatures would thaw the ground and release methane, creating a runaway greenhouse effect that would cause temperatures to rise even higher.

Many species would be unable to keep pace with these rapid climatic

(PHOTO BY J.R. WILLIAMS, COURTESY OF USGS)

Fig. 9-12. Arctic tundra in southwestern Copper River Basin, Alaska.

changes. Those species that can migrate might find their routes blocked by natural and man-made barriers. Entire biological communities would be rearranged, and many species would become extinct, while others considered pests would overrun the landscape. High levels of carbon dioxide, which acts as a fertilizer, favor the growth of weeds. It would also be a heyday for parasites and pathogens, so tropical diseases could influx into the temperate zone. The culminating effect would be a diminishing species diversity worldwide, which would also be a disaster in human terms as well.

10

Global Extinction

EXPANDING economic developments throughout the world are severely disrupting patterns of land and water use. The global destruction of forests, large-scale extraction and combustion of fossil fuels, and widespread use of man-made chemicals in industry and agriculture appear to be altering cycles of essential nutrients in the biosphere. These activities also appear to be affecting the global climate and altering precipitation patterns.

In many parts of the world, environmental protection must yield to other concerns, such as the economy—including the loss of jobs, bankrupt businesses, decreased productivity, and many other economic problems. Cleaning up the environment is an expensive undertaking, and pursuing this goal could damage the economy. This shortsightedness, however, could spell ecological disaster if dramatic steps are not taken now to remedy what will ultimately become a serious man-made disaster.

ENVIRONMENTAL POLLUTION

Humans are by far the greatest polluters in the world, and since the industrial era began, people have rivaled nature in the quantity of toxic chemicals and particulate matter disposed of in the atmosphere (FIG. 10-1). The amount of soot and dust suspended in the atmosphere at any one time as the result of human activity is estimated at about 15 million tons and this figure is increasing steadily.

Fig. 10-1. Air pollution is a serious problem resulting from industrialization.

Slash-and-burn agricultural methods clog the atmosphere with tremendous amounts of smoke. Dust entering the atmosphere from newly plowed or abandoned fields has been on the rise. Factory smokestacks discharge massive quantities of soot and aerosols. Motor vehicle exhaust alone accounts for 50 percent of the particulates and aerosols released into the atmosphere.

The hydrologic cycle (FIG. 10-2) is the Earth's cleansing agent that rids the planet of its natural and man-made pollutants by the process of dilution. However, toxic wastes that end up in the ocean from surface runoff might become concentrated to lethal levels. Even the middle of the Pacific Ocean, which was once thought to be pristine, is polluted with 100,000 particles per square mile. Many toxic substances that are diluted to supposedly safe levels in streams, lakes, and seas are concentrated by biological activities, beginning at the very bottom of the food chain (FIG. 10-3) and working on up to fish and other aquatic life. Some of this contaminated fish is a major source of food for many people throughout the world.

Each day, Americans churn out 500 thousand tons of garbage, most of which is trucked to overflowing landfills and dumped. Toxic substances leach out of the landfills and contaminate nearby water wells, requiring expensive treatment. Many toxic pollutants are powerful carcinogens and mutagens—some are nonbiodegradable and persist in the environment for long periods. Approximately eight million tons of toxic wastes are dumped into rivers and coastal waters each year (FIG. 10-4). As a result of the high cost of land disposal of toxic wastes, coastal

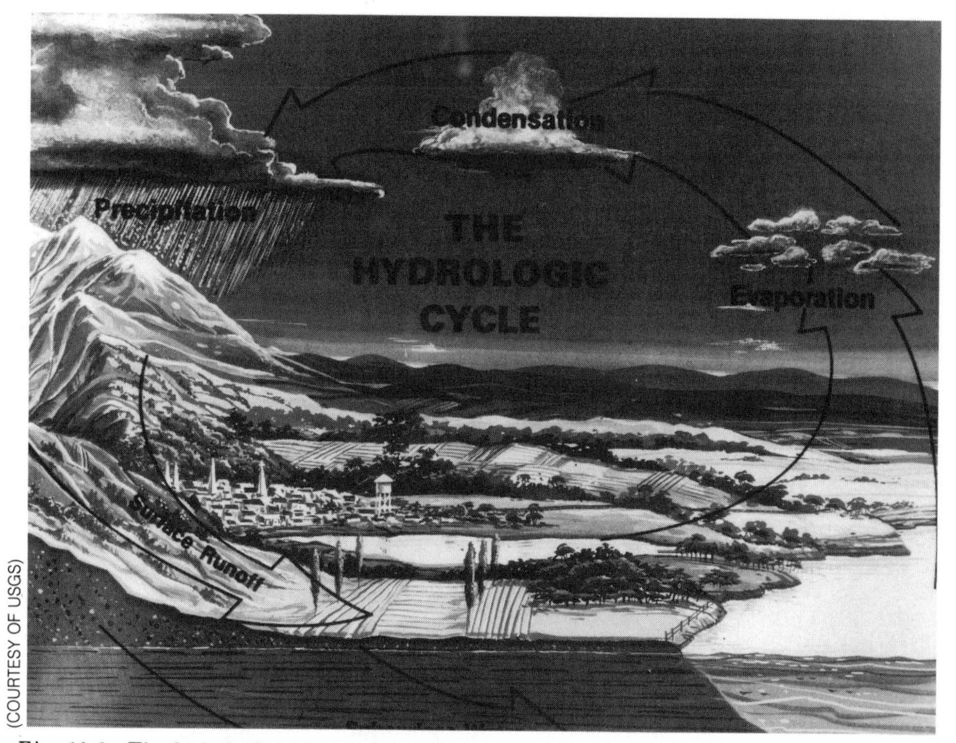

(COURTESY OF USGS)

Fig. 10-2. The hydrologic cycle involves evaporation of seawater, precipitation of rain or snow, runoff into surface streams and groundwater, and a return to the sea.

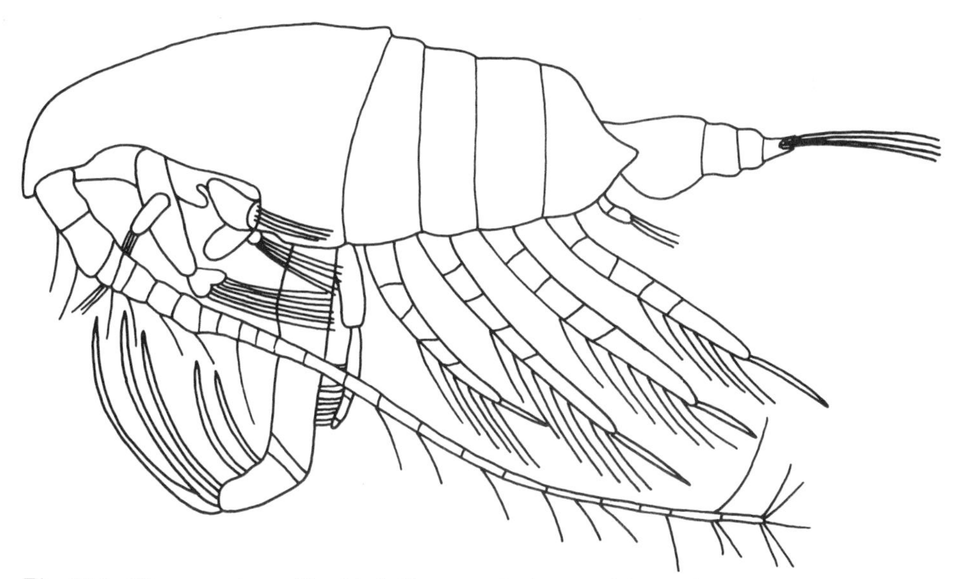

Fig. 10-3. Tiny organisms, like this krill, are at the bottom of the marine food chain and support other aquatic life.

(PHOTO BY C.H. AUBOL, COURTESY OF USDA-SOIL CONSERVATION SERVICE)

Fig. 10-4. Pollution of the Mississippi River by a broken sewage pipe near Bemidji, Minnesota.

metropolitan areas are forced to dump municipal and industrial wastes directly into the sea.

The marine dump sites are located along the East Coast; garbage barges are taken out to about 100 miles off the coast and their wastes are dumped beyond the continental shelf. Ocean currents often bring the wastes back to shore, making beaches unsightly and unsafe for swimming. Wastes are also concentrated between thermal layers and ocean fronts, wherein lie some of the most productive fishing grounds. Moreover, the meandering currents of the Atlantic Gulf Stream (FIG. 10-5), which are often laden with fish, sweep directly over the dump sites.

Of the various types of coastal pollution, nothing is as hideous as oil spills, which are on the rise. Five million tons of oil ends up in the ocean each year. Increased demand for offshore oil, collisions and groundings of oil tankers, and attacks on oil tankers by warring nations has led to disastrous ecological consequences. Heavy spills, such as the grounding of the Exxon oil tanker *Valdez* in Alaska's Prince William Sound in March 1989, often require extensive cleanup efforts (FIG. 10-6). Moreover, marine pollution does not remain localized in highly contaminated areas for very long—it eventually spreads to other parts of the world via ocean currents.

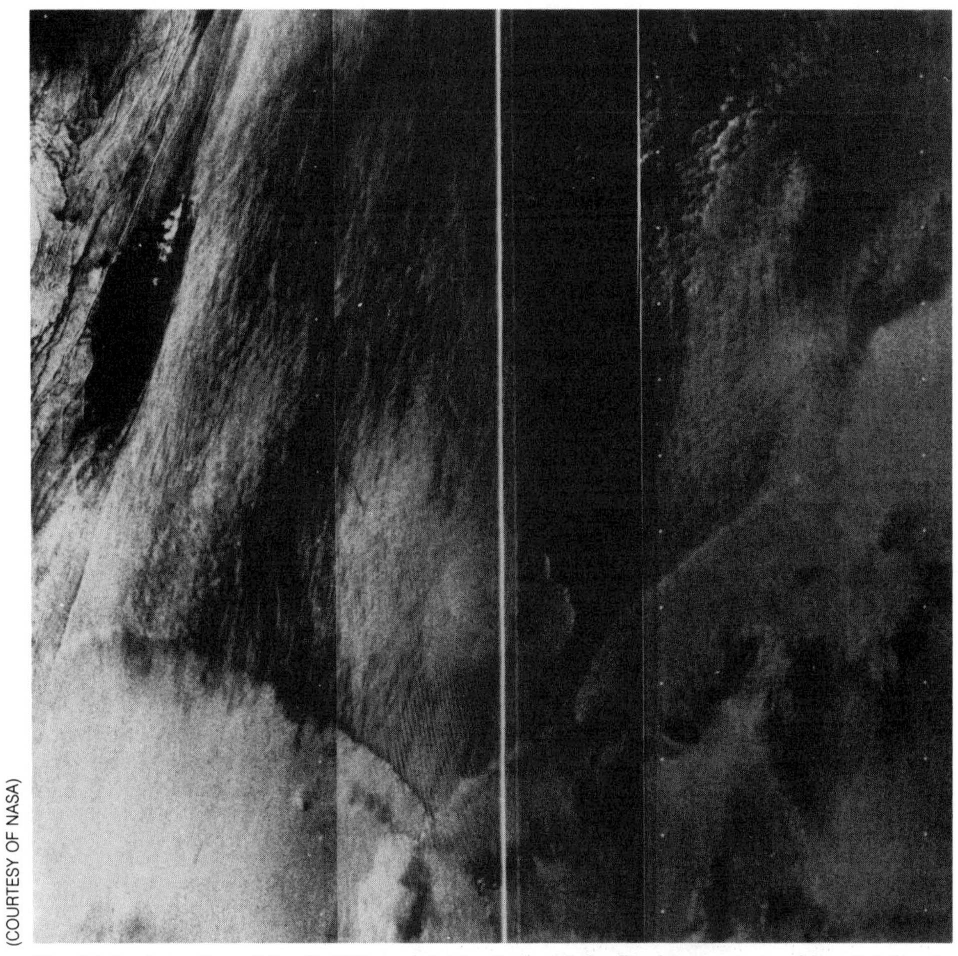

(COURTESY OF NASA)

Fig. 10-5. A portion of the Gulf Stream (upper left), off the Florida coast near Miami, taken by Seasat satellite radar, which also shows the effects of a local rain squall in the lower half of the picture.

Acid rain produced by the combustion of fossil fuels (FIG. 10-7) is especially harmful to aquatic organisms because most species cannot tolerate high acidity levels in their environment. In seawater, the damage comes from nitrogen oxides found in acid rain. Nitrogen acts like a nutrient and promotes the growth of algae. The algae blocks out sunlight and depletes the water of its dissolved oxygen, which in turn suffocates other aquatic plants and animals. In addition to widespread increases in nitrate levels, higher concentrations of toxic metals, including arsenic, cadmium, and selenium have also been noted. The main factors that contribute to this increase are fertilizer and pesticide runoffs and acid rain, which also dissolves heavy metals in the soil.

The acidity levels of rain and snow indicate that in many parts of the world, precipitation has changed from a nearly neutral solution at the beginning of the industrial era (two centuries ago) to a dilute solution of sulfuric and nitric acid. In some cases, acid rain is as strong as vinegar. Even in virtually unindustralized areas, like the tropics, acid rain occurs mostly from the burning of forests. Streams and lakes, especially those that are not buffered by carbonate rocks, which help neutralize the acid, have become so acidic from acid rain runoff or polluted by toxic wastes that many aquatic populations have been nearly decimated.

Even the soil in some areas has become so acidic it can no longer be cultivated. Plants are also damaged by the direct effects of acid on foliage and root systems. Acid precipitation is causing the destruction of the great forests of

(PHOTO BY K.A. SPANGLER, COURTESY OF U.S. COAST GUARD)

Fig. 10-6. A member of the Gulf Strike Team, which fights oil spills and toxic chemical releases.

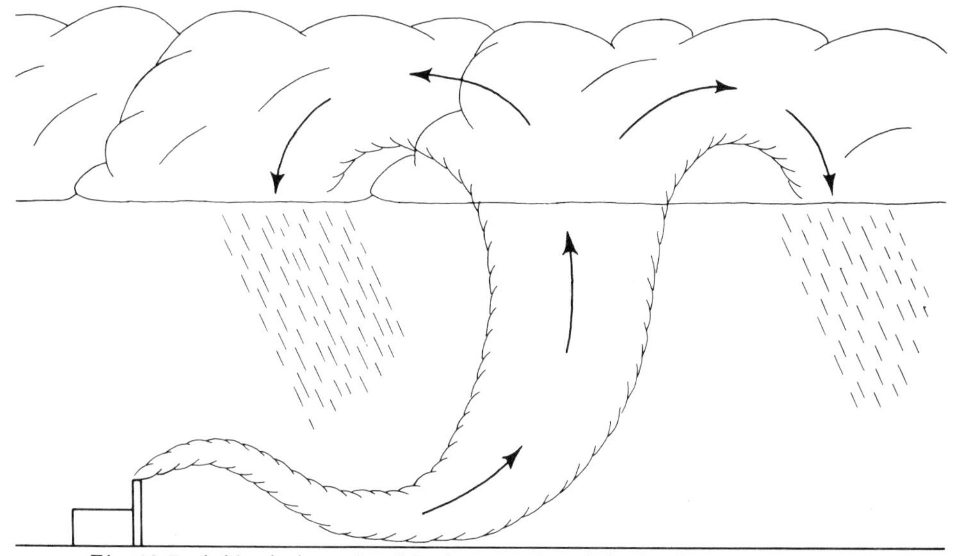

Fig. 10-7. Acid rain is produced by the scavaging of cloud pollutants by precipitation.

North America, Europe, China, and Brazil. Resorts and wilderness areas are losing much of their natural beauty as a result of acid rain.

The disposal of radioactive wastes is of great concern because of its environmental effects over very long periods and because of the proliferation of nuclear technology throughout the world. High-level radioactive wastes are among the most difficult to dispose of because of their intense nuclear radiation, high heat output, and longevity. Each disposal site requires a stable geologic formation that will prevent the contamination of the groundwater supply by leaky radioactive waste containers.

Using satellites to monitor the ozone concentration in the upper atmosphere, scientists have discovered to their horror that the ozone layer is being depleted. Every September and October since the late 1970s, a giant hole about the size of the continental United States has opened in the ozone layer over the South Pole (FIG. 10-8). A similar ozone hole has been discovered over the North Pole as well. Long-term records show that ozone levels in the high northern latitudes have dropped roughly 5 percent over the last two decades. The ozone depletion is strongly believed to have a chemical origin and these chemicals are man-made. Even if the chemical emissions stopped today, the ozone layer would continue to be depleted for at least a century, the length of time required to cleanse the stratosphere.

The ozone layer plays a vital role in shielding the Earth from harmful ultraviolet radiation from the Sun. Without this shield, life could not exist on the Earth's surface or in the surface waters of the ocean. Remember that it was the establishment of the ozone layer that allowed life to leave the ocean for a residence on land. A continued depletion of the ozone layer with accompanying high ultraviolet exposures could reduce crop productivity and aquatic life, especially primary producers, upon which most life on Earth ultimately depends for survival.

DEFORESTATION

Humans are destroying the world's forests at an alarming rate. Tropical rain forests, especially those in the Amazon Basin of South America, are decreasing by about 30 million acres annually (about the size of Alabama). The remaining tropical rain forests cover an area only about the size of the United States. The forests are cleared mostly for agriculture, but about 15 percent is cut for timber, much of which is wasted through inefficient harvesting and milling methods. Some tropical forest fires are so massive they create a gigantic smoke screen that blocks the view of the surface from space (FIG. 10-9). If global deforestation continues to escalate, most of the world's rain forests will be completely gone before the middle of the next century.

Tropical rain forests contain two-thirds or more of the world's species of plants and animals, yet they cover only 7 percent of the land surface. Species in the rain forests are being crowded out by the encroachment of humans into their habitats, along with the destruction and pollution of their ecological niches. The human race is growing so explosively and destroying the environment so extensively, that species are dying out in tragic numbers as a result.

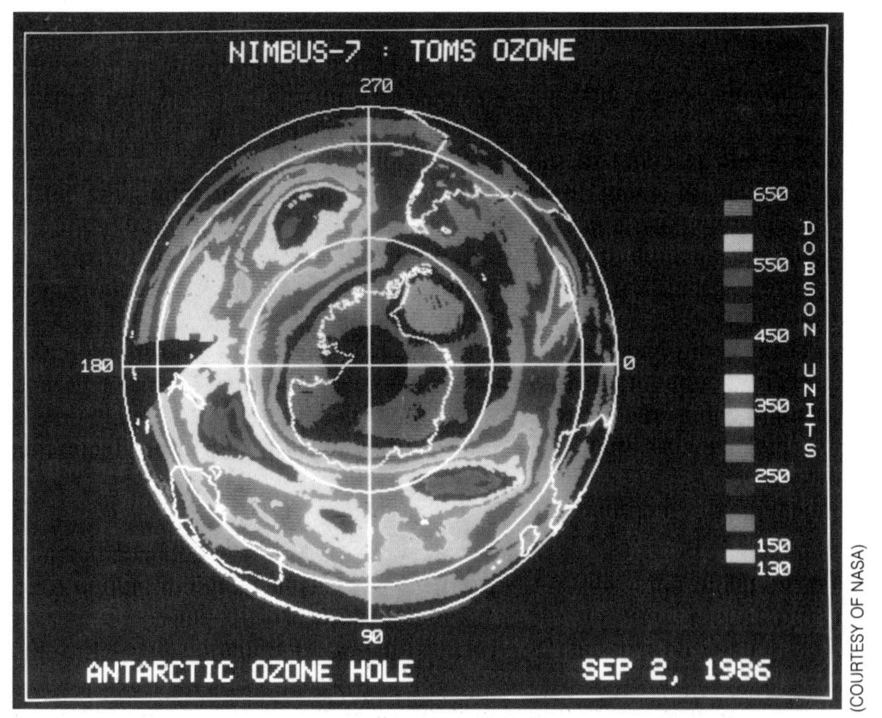

(COURTESY OF NASA)

Fig. 10-8. The Antarctic ozone hole view by Nimbus-7.

PHOTO TAKEN FROM THE SPACE SHUTTLE (COURTESY OF NASA)

Fig. 10-9. Smoke created by agricultural burning makes it impossible to view the ground at Zaire (in central Africa) from the sky

Thus far, 50 percent the forests of the world have been cut for agricultural purposes. Only about 15 percent of the once vast sea of forests in the United States still remains and these are being rapidly depleted for timber products, which are in great demand for both domestic and foreign markets. Over 80 percent of Mexico's once vast rain forest, the only one in North America, has been destroyed. Approximately 75 percent of the deforestation worldwide is conducted by landless people in a desperate search for food. Moreover, forests in Africa are being cut mostly for firewood, the only source of fuel for heating and cooking. As these forests recede, severe firewood shortages loom ahead.

Deforestation is occurring in all parts of the world, but it is most hideous in the Amazon jungle of Brazil, where 20 million acres of forests are burned each year. Thus far, about 20 percent of the Amazon rain forest has already been destroyed. The cleared land mostly provides pasture for grazing cattle, whose beef products are exported to other nations. However, after two or three years, the soil is depleted and robbed of its nutrients. Then, the land is abandoned and farmers clear more land in this highly destructive cycle.

When the rains come, flash floods wash the exposed soil down to bedrock. Without this thin covering of soil, the rain forest has no chance of recovery. Moreover, about 50 percent of the precipitation comes from the forests themselves as a result of the high transpiration rates of the dense vegetation. Because vast areas of the rain forests are being destroyed, the precipitation patterns are changing, which has the potential of turning wide areas into man-made deserts.

Modern methods of timber harvesting, including the widespread use of chainsaws, bulldozers, and giant logging machines is contributing substantially to the rapid decline of the rain forests. Some rain forests, such as those in Hawaii, are cut for electric power generation—a tragic waste of a valuable resource. After the timber companies have stripped the more desirable trees, unwanted trees and brush are burned and the bare soil, now denuded of all vegetation, is left unattended—vulnerable to flash floods and erosion.

DESERTIFICATION

The process of *desertification* is a product of the climate and human activities and causes severe degradation of the environment. It results from the loss of topsoil, taking millions of acres of once-productive agricultural land out of cultivation every year. When land loses its topsoil, only coarse, infertile sands are left behind, and a man-made desert is created. The problem worsens when the land is subjected to flash floods, higher erosion rates, and dust storms (FIG. 10-10).

Desertification is occurring over the entire world, but especially in central Africa, where the burning sands of the Sahara desert march across once fertile lands of the Sahel region to the south. The process of desertification is also self-perpetuating because the light colored sands of the desert reflect more sunlight than vegetated land. This heat produces permanent high-pressure zones that block weather systems, which normally would bring life-giving rain to the region.

The denuded land is also subjected to flash floods and dust storms, which transport the sediments out of the area and deposit them elsewhere, often where nobody wants them. Throughout the world, from 30 to 50 percent of once-fertile lands have been rendered useless by erosion and desertification (FIG. 10-11).

The act of desertification is also being played out in the tropical rain forests, which are being cleared on an unprecedented scale to make room for pastureland for grazing cattle. After a few years of extensive agriculture, the soil is weakened, and because most of the world's farmers cannot afford expensive fertilizers, the worn-out land is abandoned. Meanwhile, heavy downpours wash away the denuded top soil, often exposing bare bedrock.

The destruction of large parts of the rain forest is also changing precipitation patterns, which could be responsible for turning large areas into man-made

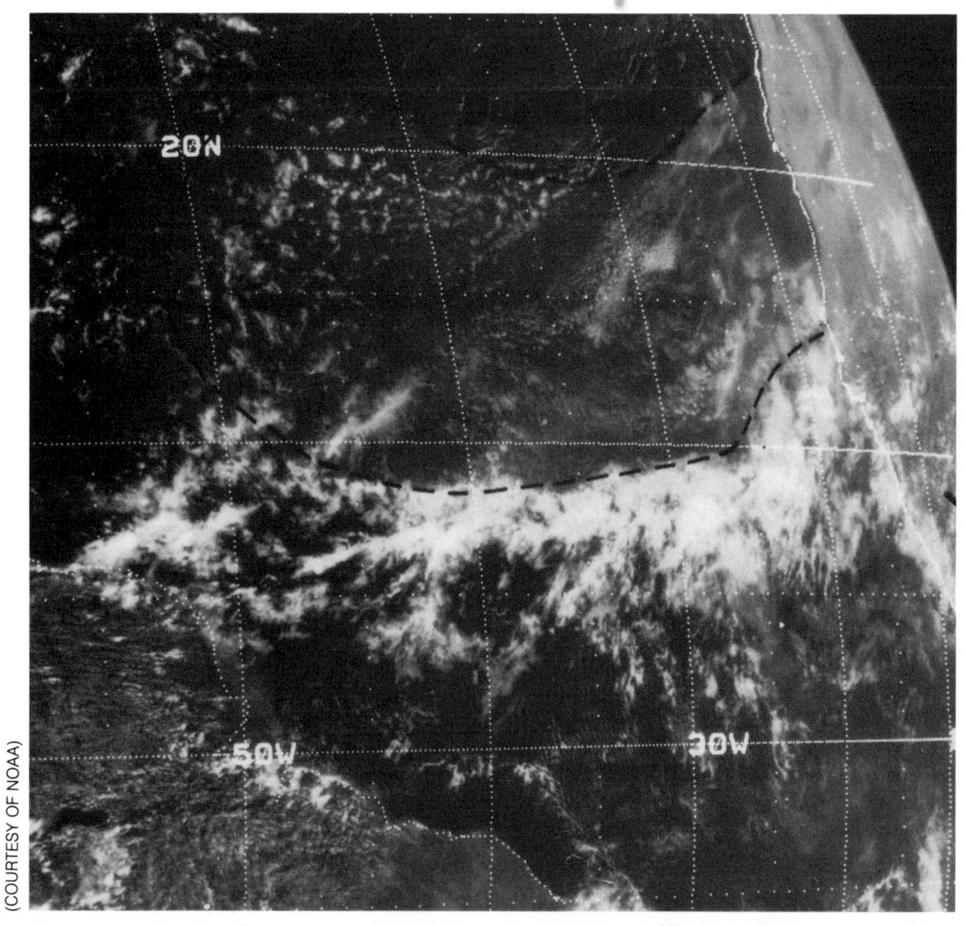

(COURTESY OF NOAA)

Fig. 10-10. During the summer of 1976, drought conditions in West Africa and a prevailing easterly wind resulted in a dust surge that created an enormous cloud of dust blowing out over the Atlantic Ocean from the Sahara Desert. This view is from the GOES 1 weather satellite.

deserts. In other deforested regions, the rainfall has actually increased, causing severe topsoil erosion, thus making forest recovery next to impossible. The denuding of the world's forests increases the reflection of solar energy, which consequently causes precipitation loss. This water loss adds an additional strain on the forests and subjects them to infestation and disease, causing further tree loss.

It also appears that the removal of the rain forests is having an adverse effect on the global climate. High evaporation and transpiration rates within the forests themselves help create clouds, which contribute to the prodigious amount of rain these areas receive. When the forests are removed, the cycle is broken. Since the Earth's climate operates as an integrated system, perturbations in one part of the world can affect the climate in other parts. As more bare areas dot the Earth's surface, the global heat budget, which stabilizes climatic conditions and allows life to exist on the planet, is bound to be detrimentally affected.

OVERPOPULATION

Human populations are growing so explosively and are modifying the environment so extensively that we are inflicting a global impact of unprecedented dimensions—possibly causing a greater havoc than the Earth has ever endured. Humans are presently consuming nearly 50 percent of the Earth's *net primary production*, the total amount of energy stored by green vegetation throughout the world. The level of human growth over this century has been staggering, and the population is expected to reach 10 billion by 2025. Humans will then need to consume twice the world's net primary production. These events could be disastrous, considering the destructive impacts of today's level of human activities.

Scientists who study human populations have concluded that the world has already reached its *carrying capacity*, the ability of the land to continue supplying people's needs. So, in the future it will be difficult to feed, clothe,

(PHOTO BY T. McCABE, COURTESY OF USDA-SOIL CONSERVATION SERVICE)

Fig. 10-11. Soil erosion is becoming a serious problem throughout the world.

shelter, and employ additional people at more than a subsistent level of life. Rapid population growth has already stretched the resources of the world, and the prospect of future increases raises serious doubts as to whether the planet can continue to support people's growing requirements. Up to a tenfold increase in world economic activity will be required over the next 50 years to keep up with basic human needs—a situation the biosphere cannot possibly tolerate without irreversible damage.

For those developing countries with the highest birth rates, economic gains are quickly eroded—simply by having too many mouths to feed. The developing nations in Asia, Africa, and South America, are in a desperate race to keep food supplies running alongside of population growth. When droughts occur and famine strikes, these people are placed in great peril, especially if for political or other reasons, food imports cannot keep up with the demand.

Nearly 50 percent of the total tonnage of crops and 75 percent of the energy and protein content is supplied by cereal grains such as rice, wheat, and barley (FIG. 10-12). Only about a dozen crops supply 90 percent of all the world's food and

(PHOTO BY T. McCABE, COURTESY OF USDA-SOIL CONSERVATION SERVICE)

Fig. 10-12. Barley that is about two weeks from harvesting.

most strains are genetically undiversified. This means that disease and infestation targeted at these specific strains could wipe out a nation's entire harvest and cause a famine of unprecedented proportions.

When agriculture can no longer supply the necessary food, the people will be in grave danger of dying from famine. During favorable climates, populations grow well beyond the limits imposed by unfavorable climates—when harvests are poor. The human race might find itself closer to the brink, when mass starvation results from severely reduced crop yields as a result of drought, infestation, or disease.

MODERN EXTINCTIONS

The oceans of the world are suffering from a dangerous decrease in vitality. The drop in marine species is much too large to be explained by chemical pollution alone, however. The decline is also caused by overfishing, diverting rivers, filling in marshes, and other destructive human activities. Moreover, riverine fisheries have been damaged by increased sedimentation resulting from the erosion and deforestation of the floodplains. The world faces a serious catastrophe if the attitude of exploiting the oceans for short-term gains does not change so that the ocean is managed properly for all inhabitants (FIG. 10-13 and (TABLE 10-1).

Fig. 10-13. The world's economic zones, where major fisheries supply human dietary needs.

Also, no particular ocean is any more or less healthy than any other; ocean currents move water from one part of the world to another, and therefore, make any differences temporary. As a result, pollutants, like DDT, have been found in the livers of penguins in the once pristine waters of the Antarctic. Pollutants in the Mediterranean Sea and other heavily polluted bodies of water, such as the Caribbean, the North Sea, the Kattegat Sea, and the Gulf of Finland will eventually

Table 10-1. Productivity of the Oceans

Location	Primary Production Tons Per Year of Organic Carbon	Percent	Total Available Fish Tons Per Year of Fresh Fish	Percent
Oceanic	16.3 billion	81.5	.16 million	0.07
Coastal Seas	3.6 billion	18.0	120.00 million	49.97
Upwelling Areas	0.1 billion	0.5	120.00 million	49.97
Total	20.0 billion		240.16 million	

pollute the entire ocean. Rivers and enclosed or semienclosed seas are presently in worse shape than the open ocean. That situation, however, is rapidly changing, and the ocean is in danger of becoming equally polluted.

The extinctions of the past were caused by natural phenomena, such as changing climatic conditions. The present extinction rate is on the order of 10,000 times greater than it was prior to the appearance of human beings. However, present-day extinctions, through which an astonishing rate of four species are vanishing every minute, are forced extinctions caused by destructive human activities. If the present spiral of human population and environmental destruction continues out of control, possibly by the middle of the next century, 50 percent or more of the species living on Earth today will be gone. Already, many species nearing extinction can only be found in captivity. These animals are "the cute and cuddlies" of the environmental movement and only represent a very small fraction of all endangered species.

Man is the only creature on Earth that is affecting the extinctions of large numbers of other species. All plants and animals not directly beneficial to humans will be forced aside as growing human populations continue to squander the Earth's space and resources and contaminate the environment. Scientists still do not fully comprehend the complex interrelationships among species and between them and their environments. However, it is becoming more apparent that the destruction of large numbers of species will leave the biological world entirely different.

Large herbivors, like the African elephant and black rhinoceros, the latter of which has existed for more than 70 million years, are in danger of becoming extinct as a result of a greedy ivory trade and human encroachment upon their environment. Only a few decades ago, Africa was a sea of wild animals surrounding a few islands of humans. Today. however, that situation is entirely reversed.

The African birth rate is the highest in the world, due to triple to 1.6 billion by 2025. The large herbivors have the capability of actually improving their environment. They open forests for grasslands, which increases productivity and accelerates nutrient recycling. Unfortunately, when the animals are eliminated, these favorable environmental impacts are reversed and the habitats of smaller herbivors are restricted.

Some exotic plants that are quickly becoming extinct have important medicinal values. Over 50 percent of the pharmaceuticals manufactured are derived from natural herbs, most of which exist only in the rain forests. Unfortunately, plants receive relatively little attention and few conservation funds. In the United States alone, as much as 10 percent of the nation's 25,000 species of plants are destined for extinction by the end of the 20th century. The native plants are at risk of extinction from habitat loss as a result the destruction of forests, the extension of agriculture, and urban sprawl. For the entire world, about 7 percent of all plant species will probably go extinct by the end of the century if the current trends continue.

Tropical rain forests in the New World (FIG. 10-14) once covered an area of 3 million square miles. Today, that area has reduced by about 30 percent. The tropical rain forests of Africa have been reduced by as much as 75 percent since 1960. Some nations, aware of the bleak future that awaits many species, are setting aside forested areas in an attempt to halt the tide of deforestation and extinction—and perhaps preserve the Earth for all creatures great and small.

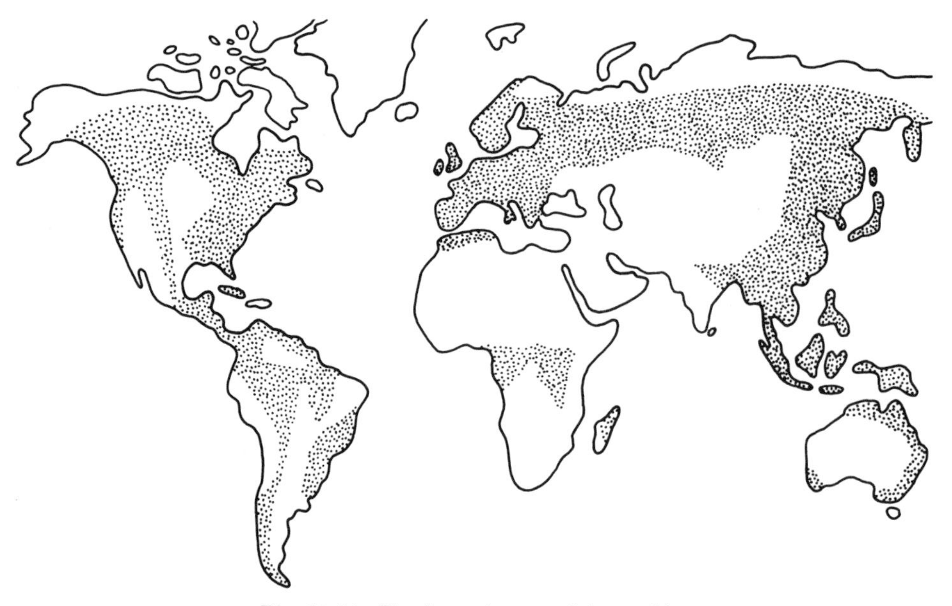

Fig. 10-14. The forested areas of the world.

THE WORLD AFTER

If environmental stress produces a new species, then humans can be considered products of the ice ages. The factors that contributed to human success were not the same as those that favored the mammals over the dinosaurs, however. For humans have one quality that no other animal possesses, the ability to alter his environment to suit his purposes. He is also the most adaptable animal on Earth, able to exist in nearly all environments. This high degree of adaptability is the hallmark of human success.

As long as human populations remained small and the environmental impacts of human activities were negligible, the rest of the world was virtually unaffected. However, after the invention of agriculture, the human character changed, and populations began to soar. As a result, the pressure of human presence was felt by the rest of the living world.

Sometime during the next century, if the present rate of extinction continues, the number of species lost as a result of human activities could surpass that of the great die-out at the end of the Cretaceous period, when the dinosaurs and 70 percent of all other species disappeared. With our burgeoning human populations

(COURTESY OF NASA)

Fig. 10-15. A view of the Earth from Apollo 17 on its way back from the Moon. Either we can continue to wantonly pollute the environment and destroy wild life habitat or we can preserve the planet for the benefit of all living things—the choice is ours.

and high levels of habitat destruction and pollution, other species are forced out of the competition, allowing the rise of more hardy species, some of which might be very destructive and harmful. In other words, by destroying predators, those species commonly called "pests" flourish, thereby upsetting the balance of nature.

We certainly have been blessed with a highly diverse and interesting biosphere, filled with as many species as have ever lived at one time. It would be a great tragedy if through our neglect of the environment and wanton destruction of life, we were to so perturb the planet as to return it to a condition of low diversity by destroying vast numbers of species. We would then have to wait for perhaps millions of years before the Earth would recover from our folly. We could instead work toward preserving the only known living planet in the entire universe (FIG. 10-15).

Glossary

abyssal—The deep ocean, generally over a mile deep.

aerosol—A mass, made of solid or liquid particles dispersed in air.

age—A geological time interval that is smaller than an epoch.

albedo—The amount of sunlight reflected from an object.

alluvium—Stream-deposited sediment.

alpine glacier—A mountain glacier or a glacier in a mountain valley.

anticline—Folded sediments that slope downward from a central axis.

aquifer—A subterranian bed of sediments, through which groundwater flows.

ash, volcanic—Fine pyroclastic material injected into the atmosphere by an erupting volcano.

asteroid—A rocky or metallic body, orbiting the Sun between Mars and Jupiter.

asthenosphere—A layer of the upper mantle, roughly between 50 and 200 miles below the surface, and is more plastic than the rock above and below and might be in convective motion.

atmospheric pressure—The weight per unit area of the total mass of air above a given point; also called *barometric pressure.*

aurora—Luminous bands of colored light seen near the poles, caused by cosmic ray bombardment of the upper atmosphere.

basalt—A volcanic rock that is dark in color and usually quite fluid in its molten state.

big bang—A theory for the creation of the universe, wherein all matter started with a singularity of infinite density and infinitesimally small size.

biogenic—Sediments composed of the remains of plant and animal life, such as shells.

biosphere—The living portion of the Earth that interacts with all other geological and biological processes.

black smoker—Superheated hydrothermal water rising to the surface at a midocean ridge. The water is supersaturated with metals, and when exiting through the sea floor, the water quickly cools and the dissolved metals precipitate, resulting in black, smokelike effluent.

caldera—A large pitlike depression found at the summits of some volcanoes that is formed by great explosive activity and collapse.

calving—Icebergs that form when glaciers break upon entering the ocean.

Cambrian explosion—Relating to a rapid radiation of species after the close of the Precambrian. This "explosion" occurred as a result of a large adaptive space, including a large number of habitats and a mild climate.

carbonaceous—A substance containing carbon, namely sedimentary rocks, such as limestone and certain types of meteorites.

carbonate—A mineral containing calcium carbonate, such as limestone and dolostone.

Cenozoic—The era of geologic time that comprises the last 65 million years.

conduction—The transmission of energy through a medium.

conglomerate—A sedimentary rock composed of welded fine-grained and coarse-grained rock fragments.

continent—A slab of light, granitic rock that floats on the denser rocks of the upper mantle.

continental drift—The concept that the continents have been drifting across the surface of the Earth throughout geologic time.

continental glacier—An ice sheet that covers a portion of a continent.

continental shelf—The offshore area of a continent in shallow sea.

continental shield—Ancient crustal rocks upon which the continents grew.

continental slope—The transition between the continental margin and the deep sea basin.

convection—A circular, vertical flow of a fluid medium as a result of heating from below. As materials are heated, they become less dense and rise; cooler, heavier materials sink.

coral—Any of a large group of shallow-water, bottom-dwelling marine invertebrates that commonly build reef colonies in warm waters.

core—The central part of a planet that consists of a heavy iron-nickel alloy.

Coriolis effect—The apparent force that deflects wind and ocean currents, and causes them to curve in relation to the rotating Earth.

correlation—To trace equivalent rock exposures over a distance, usually with the aid of fossil beds.

cosmic rays—High-energy charged particles that enter the Earth's atmosphere from outer space.

crater, meteoritic—A depression in the crust produced by a meteorite fall.

crater, volcanic—The inverted conical depression found at the summit of most volcanoes and formed by the explosive emission of volcanic ejecta.

craton—The stable interior of a continent, usually composed of the oldest rocks on the continent.

Cretaceous—The period of geologic time that encompasses from 135 to 65 million years ago.

crust—The outer layers of rocks on a planet or moon.

crustal plate—One of several plates that comprise the Earth's surface rocks.

diapir—The buoyant rise of molten rock through heavier rock.

drought—A period of abnormally dry weather that is sufficiently prolonged by the lack of water. Droughts cause serious deleterious effects on agricultural and other biological activities.

dune—A ridge of wind-blown sediments that are usually in motion.

dynamo effect—The creation of the Earth's

magnetic field by rotational, thermal, chemical, and electrical differences between the solid inner core and the liquid outer core.

earthquake—The volcanic or tectonic upheaval of the Earth's crust.

East Pacific Rise—A midocean spreading center that runs north-south along the eastern side of the Pacific. The predominant location upon which the hot springs and black smokers have been discovered.

electromagnetic radiation—Energy from the Sun that travels through the vacuum of space to reach the Earth as electromagnetic waves.

electron—A negative particle of small mass that orbits the nucleus and is equal to the number of protons.

element—A material that consists of only one type of atom.

eolian—A deposit of wind-blown sediment.

epoch—A geologic time unit that is shorter than a *period* and longer than an *age.*

equinox—Either of the two points of intersection of the Sun's path and the plane of the Earth's equator.

estuary—A tidal inlet along a coast that contains abundant species.

evaporation—The transformation of a liquid into a gas.

evolution—The tendency of physical and biological factors to change with time.

extraterrestrial—That which pertains to all phenomena outside the Earth.

extrusive—Any igneous volcanic rock that is ejected onto the surface of a planet or moon.

fault—A breaking of crustal rocks caused by earth movements.

fissure—A large crack in the crust through which magma might escape from a volcano.

foraminifera—Calcium carbonate secreting organisms that live in the surface waters of the oceans; after death, their shells are deposited on the sea floor where they form the primary constituents of limestone.

formation—A combination of rock units that can be traced over distance.

fossil—Any remains or impression in rock of a plant or animal of a previous geologic age.

glacier—A mass of moving ice.

glossopteris—A late Paleozoic plant that existed on the southern continents, however it has not been found on the northern continents. This evidence confirms the existence of Gondwana.

Gondwana—A southern supercontinent of Paleozoic time, consisting of Africa, South America, India, Australia, and Antarctica. It broke into the present continents during the Mesozoic era.

granite—A coarse-grained, silica-rich rock that consists primarily of quartz and feldspars. It is the principal constituent of the continents and is believed to be derived from a molten state beneath the Earth's surface.

greenhouse effect—The trapping of heat in the atmosphere, principally by water vapor and carbon dioxide.

groundwater—The water derived from the atmosphere that percolates and circulates below the surface of the Earth.

half-life—The time for one-half the atoms of a radioactive element to decay.

helium—The second lightest and second most abundant element in the universe; helium is composed of two protons and two neutrons.

Holocene—A geological time period that covers the last 10,000 years.

hot spot—A volcanic center that has no relation to plate boundary location; an anomalous magma generation site in the mantle.

hydrocarbon—A molecule that consists of carbon chains with attached hydrogen atoms.

hydrogen—The lightest and most abundant element in the universe; hydrogen is composed of one proton and one electron.

hydrosphere—The water layer at the surface of the Earth.

hydrothermal—Relating to the movement of hot water through the crust.

ice age—A period of time when large areas of the Earth were covered by glaciers.
iceberg—A portion of a glacier that breaks off upon entering the sea.
ice cap—A polar cover of ice and snow.
igneous rocks—All rocks that have solidified from a molten state.
impact—The point on the surface upon which a celestial object lands.
infrared—Heat radiation with a wavelength between red light and radio waves.
insolation—All solar radiation that impinges on a planet.
interglacial—A warming period between glacial periods.
intrusive—Any igneous body that has solidified in place below the surface of the Earth.
ionization—The process whereby electrons are torn from previously neutral atoms.
ionosphere—The atmospheric shell that is characterized by high ion density; it extends from about 40 miles up to very high regions of the atmosphere.
iridium—A rare isotope of platinum that is relatively abundant on meteorites.
isotope—A derivative of particular element that has a different number of neutrons in the nucleus.

jet stream—Relatively strong winds concentrated within a narrow belt that is usually found in the tropopause.

Laurasia—The northern supercontinent of the Paleozoic era that consisted of North America and Eurasia.
lava—Molten magma after it has flowed onto the surface.
light-year—The distance light travels in one year—about six trillion miles.
limestone—A sedimentary rock that consists mostly of calcite.
lithosphere—The rigid outer layer of the mantle, which is typically about 60 miles thick. It is overridden by the continental and oceanic crusts and is divided into segments, called *plates.*

magma—A molten rock material that is generated within the Earth and is the constituent of igneous rocks.
magnetic field reversal—A reversal of the north-south polarity of a planet's magnetic poles.
magnetosphere—The region of the Earth's upper atmosphere in which the Earth's magnetic field controls the motion of ionized particles.
mantle—The part of a planet that is below the crust and above the core, and is composed of dense iron magnesium-rich rocks.
maria—Dark plains on the lunar surface that are caused by massive basalt floods.
Mesozoic—Literally the period of middle life; it refers to the period between 240 and 65 million years ago. The Mesozoic is also synonymous with the age of the dinosaurs.
meteorite—A metallic or stony body from space that enters the Earth's atmosphere and impacts on the Earth's surface.
methane—A hydrocarbon gas that is liberated by the decomposition of organic matter.
mid-Atlantic ridge—The seafloor spreading ridge of volcanoes that marks the extensional edge of the North and South American plates to the west and the Eurasian and African plates to the east.
midocean ridge—A submarine ridge along a divergent plate boundary, where a new ocean floor is created by the upwelling of mantle material.
monsoon—A seasonal wind that accompanies temperature changes over land and water from one season of the year to another.

nebula—An extended astronomical object with a cloudlike appearance. Some nebulae are galaxies; others are clouds of dust and gas within our galaxy.

Nemesis—The hypothetical sister star of the Sun, which might be responsible for disturbing comets in the Oort Cloud and causing them to fall into the inner Solar System.

neutron—A particle with no electrical charge that has roughly the same weight as the positively charged proton, both of which are found in the nucleus of an atom.

nutrient—A food substance that nourishes living organisms.

Oort Cloud—The collection of comets that surround the Sun about a light-year away.

orogeny—A process of mountain building caused by tectonic activity.

outgassing—The loss of gas within a planet as opposed to *degassing*, the loss of gas from meteorites.

ozone—A molecule consisting of three atoms of oxygen that exists in the upper atmosphere and filters out ultraviolet light from the Sun.

paleomagnetism—The study of the Earth's magnetic field, including the position and polarity of the poles in the past.

paleontology—The study of ancient life-forms, based on the fossil record or plants and animals.

Paleozoic—The period of ancient life, between 570 and 240 million years ago.

Pangaea—An ancient supercontinent that included all of the Earth's landmasses.

Panthalassa—The great world ocean that surrounded Pangaea.

permafrost—Permanently frozen ground.

photon—A packet of electromagnetic energy, that is generally considered to be a particle.

photosynthesis—The process by which plants create carbohydrates from carbon dioxide, water, and sunlight.

planetesimal—Small celestial bodies that might have existed in the early stage of the Solar System.

plate tectonics—The theory that accounts for the major features of the Earth's surface in terms of the interaction of crustal plates.

polar wandering—The movement of the geographic poles.

prebiotic—Conditions, such as those on Earth, prior to the introduction of life processes.

precipitation—Products of condensation that fall from clouds as rain, snow, hail, or drizzle.

primordial—Pertaining to the primitive conditions that exist during early stages of development.

proton—A large particle with a positive charge in the nucleus of an atom.

quartz—A common igneous rock-forming mineral of silicon dioxide.

radiation—The process by which energy from the Sun is propagated through a vacuum of space as electromagnetic waves. A method, along with conduction and convection, of transporting heat.

radioactivity—An atomic reaction that releases detectable radioactive particles.

radiometric dating—To determine how long an object has existed by chemically analyzing the stable verses unstable radioactive elements.

reef—The biological community that lives at the edge of an island or continent. The shells form a limestone deposit that is readily preserved in the geologic record.

regression—A drop in sea level, that exposes continental shelves to erosion.

reversed magnetism—A geomagnetic field with the opposite polarity of the present one.

rift valley—The center of an extensional spreading center, where continental or oceanic plate separation occurs.

sea-floor spreading—The theory that the ocean floor is created by the separation of crustal plates along the midocean ridges, with new oceanic crust formed from mantle material that rises from the mantle to fill the rift.

sedimentary rock—A rock composed of sentiment that is cemented together.

shield—Areas of the exposed Precambrian nucleus of a continent.

shield volcano—A broad, low-lying volcanic cone built up by lava flows of low viscosity.

solar flare—A short-lived bright event on the Sun's surface that causes greater ionization of the Earth's upper atmosphere as a result of an increase in ultraviolet light.

solar wind—An outflow of particles from the Sun that represents the expansion of the corona.

storm surge—An abnormal rise of the water level along a shore as a result of wind flow in a storm.

stratosphere—The upper atmosphere above the troposphere, about ten miles above sea level.

stromatolite—A calcareous structure built by successive layers of bacteria that have been in existence for the past 3.5 billion years.

subduction zone—An area where an oceanic plate dives below a continental plate into the mantle. Ocean trenches are the surface expression of a subduction zone.

sunspot—A region on the Sun's surface that is cooler than surrounding regions. Sunspots affect radio transmissions on Earth.

supernova—An enormous stellar explosion in which all but the inner core of a star is blown into interstellar space, producing as much energy in a few days as the Sun does in a billion years.

surge glacier—A continental glacier that heads toward the sea at a high rate of advance.

tectonic activity—The formation of the Earth's crust by large-scale earth movements throughout geologic time.

terrestrial—All phenomena that pertains to the Earth.

Tethys Sea—The hypothetical mid-latitude area of the oceans that separated the northern and southern continents of Gondwana and Laurasia around 300 million years ago.

tide—A bulge in the ocean produced by the Moon's gravitational forces on the Earth's oceans. The rotation of the Earth beneath this bulge causes the sea level to rise and lower.

transgression—A rise in sea level that causes flooding of the shallow edges of continental margins.

troposphere—The lowest 9 to 12 miles of the Earth's atmosphere, which is characterized by temperatures that decrease with height.

tundra—Ground that is permanently frozen at high latitudes and high altitudes.

ultraviolet—The invisible light with a wavelength shorter than visible light and longer than X-rays. Ultraviolet light is harmful to exposed organisms.

volcanism—Any type of volcanic activity.

volcano—A fissure or vent in the crust through which molten rock rises to the surface.

X-rays—Electromagnetic radiation of high-energy, with wavelengths that are above the ultraviolet and below gamma rays.

Bibliography

THE ORIGIN OF LIFE

Cairns-Smith, A. G. "The First Organisms." *Scientific American* Vol. 252 (June 1985): 90–100.

Groves, David I., John S. R. Dunlop, and Roger Buick. "An Early Habitat of Life." *Scientific American* Vol. 245 (October 1981): 64–73.

Kasting, James F., Owen B. Toon, and James B. Pollack. "How Climate Evolved on the Terrestrial Planets." *Scientific American* Vol. 258 (February 1988): 90–97.

Lewin, Roger. "RNA Catalysis Gives Fresh Perspective on the Origin of Life." *Science* Vol. 231 (February 7, 1986): 545–546.

Lunine, J. I. "Origin and Evolution of Outer Solar System Atmospheres." *Science* Vol. 245 (July 14, 1989): 141–146.

Moss, I. G. "How Did the Universe begin?" *Nature* Vol. 316 (August 8, 1985): 482–483.

Raloff, Janet. "Clues to Life's Cellular Origins." *Science News* Vol. 130 (August 2, 1988): 71.

Towe, Kenneth M. "Earth's Early Atmosphere." *Science* Vol. 235 (January 23, 1987): 415.

Waldrop, M. Mitchell. "Spontaneous Order, Evolution, and Life." *Science* Vol. 247 (March 30, 1990): 1543–1545.

Wharton, Robert A., Jr. "Gathering Evidence: The Case for Past Life on Mars." *Space World* Vol. 298 (September 1988): 20–24.

THE HISTORY OF LIFE

Bower, Bruce. "Fossils Flesh Out Early Vertebrates." *Science News* Vol. 133 (January 9, 1988): 21.

Brock, Thomas D. "Precambrian Evolution." *Nature* Vol. 288 (November 20, 1980): 214–215.

Cloud, Preston. "The Biosphere." *Scientific American* Vol. 249 (September 1983): 176–189.

Conway, Simon and H. B. Whittington. "The Animals of the Burgess Shale." *Scientific American* Vol. 241 (July 1979): 122–133.

Ford, Trevor D. "Life in the Precambrian." *Nature* Vol. 285 (May 22, 1980): 193–194.

McMenamin, Mark A. S. "The Emergence of Animals." *Scientific American* Vol. 256 (April 1987): 94–102.

Morris, S. Conway. "Burgess Shale Faunas and the Cambrian Explosion." *Science* Vol. 246 (October 20, 1989): 339–345.

Schopf, William J. and Bonnie M. Parker. "Early Archean Microfossils from Warrawoona Group, Australia." *Science* Vol. 237 (July 3, 1987): 70–72.

Valentine, James W. and Eldridge M. Moores. "Plate Tectonics and the History of Life in the Oceans." *Scientific American* Vol. 230 (April 1974): 80–89.

Weisburd, Stefi. "The Microbes that Loved the Sun." *Science News* Vol. 129 (February 15, 1986): 108–110.

THE MAJOR EXTINCTIONS

Bower, Bruce. "A 'Mosaic' Ape Takes Shape." *Science News* Vol. 127 (January 12, 1985): 26–27.

Diamond Jared M. "Extinctions, Catastrophic and Gradual." *Nature* Vol. 304 (August 4, 1983): 396–397.

Lewin, Roger. "Modern Human Origins Under Close Scrutiny." *Science* Vol. 239 (March 11, 1988): 1240–1241.

______. "Extinctions and the History of Life." *Science* Vol. 221 (September 2, 1983): 935–937.

Monastersky, Richard. "Abrupt Extinctions at End of Triassic." *Science News* Vol. 132 (September 5, 1987): 149.

Raup, David M. "Biological Extinctions in Earth History." *Science* Vol. 231 (March 28, 1986): 1528–1533.

Russell, Dale A. "The Mass Extinctions of the Late Mesozoic." *Scientific American* Vol. 256 (January 1982): 58–65.

Ward, Peter. "The Extinctions of the Ammonites." *Scientific American* Vol. 249 (October 1983): 136–147.

Weisburd, Stefi. "Extinction Wars." *Science News* Vol. 129 (February 1, 1986): 75–77.

Wellnhofer, Peter. "Archaeopteryx." *Scientific American* Vol. 262 (May 1990): 70–77.

THE EVOLUTION OF SPECIES

Bakker, Robert T. "Evolution by Revolution." *Science 85* Vol. 6 (November 1985): 72–81.

Dickerson, Richard E. "Cytochrome C and the Evolution of Energy Metabolism." *Scientific American* Vol. 242 (March 1980): 137–153.

Gould, Stephen J. "Darwinism Defined: The Difference Between Fact and Theory." *Discover* Vol. 8 (January 1987): 64–70.

Herbert, Sandra. "Darwin as a Geologist." *Scientific American* Vol. 254 (May 1986): 116–123.

Kerr, Richard A. "No Longer Willful, Gaia Becomes Respectful." *Science* Vol. 240 (April 22, 1988): 393–395.

Lewin, Roger. "A Lopsided Look at Evolution." *Science* Vol. 241 (July 15, 1988): 291–293.

______. "Statistical Traps Lurk in the Fossil Record." *Science* Vol. 236 (May 1, 1987): 521–522.

Monastersky, Richard. "The Plankton-Climate Connection." *Science News* Vol. 132 (December 5, 1987): 362–354.

______. "Gaia: The Life of a Theory." *Science News* Vol. 132 (December 5, 1987): 364.

Ridley, Mark. "Evolution and Gaps in the Fossil Record." *Nature* Vol. 286 (July 31, 1980): 444–445.

Stebbins, G. Ledyard and Francisco J. Ayala. "The Evolution of Darwinism." *Scientific American* Vol. 253 (July 1985): 72–82.

THE EFFECTS OF EXTINCTIONS

Benton, Michael J. "Interpretations of Mass Extinctions." *Nature* Vol. 314 (April 11, 1985): 496–497.

Hendricks, Melissa. "Experiments Challenge Genetic Theory." *Science News* Vol. 134 (September 10, 1988): 166.

Jablonski, David. "Background and Mass Extinctions: The Alternation of Macroevolutionary Regimes." *Science* Vol. 231 (January 10, 1986): 129–133.

Lewin, Roger. "Biologists Disagree Over Bold Signature of Nature." *Science* Vol. 244 (May 5, 1989): 527–528.

______. "Mass Extinctions Select Different Victims." *Science* Vol. 231 (January 17, 1986): 219–220.

Monastersky, Richard. "Periodic Mass Extinction at Random." *Science News* Vol. 132 (November 14, 1987): 219.

Mossman, David J. and William A. S. Sarjeant. "The Footprints of Extinct Animals." *Scientific American* Vol. 248 (January 1983): 75–85.

Sepkoski, J. John, Jr. "Environmental Trends in Extinction During the Paleozoic." *Science* Vol. 235 (January 2, 1987): 64–65.

CELESTIAL CAUSES OF EXTINCTION

Grieve, Richard A. F. "Impact Cratering on the Earth." *Scientific American* Vol. 262 (April 1990): 66–73.

Kerr, Richard A. "Snowbird II: Clues to Earth's Impact History." *Science* Vol. 242 (December 9, 1988): 1380–1382.

______. "Huge Impact Is Favored K-T Boundary Killer." *Science* Vol. 242 (November 11, 1988): 865–867.

Luis W. Alvarez. "Mass Extinctions Caused by Large Bolide impacts." *Physics Today* Vol. 40 (July 1987): 24–33.

Monastersky, Richard. "Microbes Complicate the K-T Mystery." *Science News* Vol. 136 (November 25, 1989): 341.

Paresce, Francesco and Stuart Bowyer. "The Sun and the Interstellar Medium." *Scientific American* Vol. 255 (September 1986): 93–99.

Raup, David M. and J. John Sepkoski, Jr. "Periodic Extinctions of Families and Genera." *Science* Vol. 231 (February 21, 1986).

Schwarzschild, Bertram. "Do Asteroid Impacts Trigger Geomagnetic Reversals?" *Physics Today* Vol. 40 (February 1987): 17–20.

Simon, Cheryl. "Death Star." *Science News* Vol. 125 (April 21, 1984): 250–252.

Thomsen, D.E. "Mass Extinctions: Galactic Yo-Yo Effect." *Science News* Vol. 125 (June 23, 1984): 388–389.

Wolfendale, Arnold. "A Supernova for a Neighbor?" *Nature* Vol. 319 (January 9, 1986): 99.

Waldrop, M. Mitchel. "After the Fall." *Science* Vol. 239 (February 26, 1988): 977.

TERRESTRIAL CAUSES OF EXTINCTION

Bloxham, Jeremy and David Gubbins. "The Evolution of the Earth's Magnetic Field." *Scientific American* Vol. 261 (December 1989): 68–75.

Fisher, Arthur. "What Flips the Earth's Field." *Popular Science* Vol 232 (January 1988): 71–74 and 112.

Hallam, Anthony. "End-Cretaceous Mass Extinction Event: Argument for Terrestrial Causation." *Science* Vol. 238 (November 27, 1987): 1237–1241.

Hoffman, Kenneth A. "Ancient Magnetic Reversals: Clues to the Geodynamo." *Scientific American* Vol. 258 (May 1988): 76–83.

Kerr, Richard A. "Take Your Choice: Ice Ages, Quakes, or Impacts." *Science* Vol. 243 (January 27, 1989): 479–480.

______. "Was There a Prelude to the Dinosaurs' Demise?" *Science* Vol. 239 (February 12, 1988): 729–730.

Monastersky, Richard. "The Whole-Earth Syndrome." *Science News* Vol. 133 (June 11, 1988): 378–380.

Nance, R. Damian, Thomas R. Worsley, and Judith B. Moody. "The Supercontinent Cycle." *Scientific American* Vol. 259 (July 1988): 72–79.

Officer, Charles B. and Charles L. Drake. "The Cretaceous-Tertiary Transition." *Science* Vol. 219 (March 25, 1983): 1383–1390.

Rampino, Michael R. and Richard B. Strothers. "Flood Basalt Volcanism During the Past 250 Million Years." *Science* Vol. 241 (August 5, 1988): 663–667.

Weisburd, Stefi. "Volcanoes and Extinctions: Round Two." *Science News* Vol. 131 (April 18, 1987): 248–250.

White, Robert S. and Dan P. McKenzie. "Volcanism at Rifts." *Scientific American* Vol. 261 (July 1989): 62–71.

THE ICE AGES

Bower, Bruce. "Extinctions on Ice." *Science News* Vol. 132 (October 31, 1987): 284–285.

Broecker, Wallace S. and George H. Denton. "What Drives Glacial Cycles?" *Scientific American* Vol. 262 (January 1990): 49–56.

Crowley, Thomas J. and Gerald R. North. "Abrupt Climate Change and Extinction Events in Earth History." *Science* Vol. 240 (May 20, 1980): 996–1001.

Kerr, Richard A. "Did the Roof of the World Start an Ice Age." *Science News* Vol. 244 (June 23, 1989): 1441–1442.

______. "Domino Effect Invoked in Ice Age Extinctions." *Science* Vol. 238 (December 11, 1987): 1509–1510.

Lewin, Roger. "A Thermal Filter to Extinction." *Science* Vol. 223 (January 27, 1984): 383–385.

______. "Domino Effect Invoked in Ice Age Extinctions." *Science* Vol. 238 (December 11, 1987): 1509–1510.

Matthews, Samuel W. "Ice on the World." *National Geographic* Vol. 171 (January 1987): 84–103.

Monastersky, Richard. "Warm Cretaceous Earth: Don't Hold the Ice." *Science News* Vol. 133 (June 18, 1988): 391.

Overpeck, Jonathan T., et al. "Climate Change in the Circum-Atlantic Region During Last Deglaciation." *Nature* Vol. 338 (April 13, 1989): 553–556.

Stanley, Steven M. "Mass Extinctions in the Ocean." *Scientific American* Vol. 250 (June 1984): 64–72.

GREENHOUSE WARMING

Berner, Robert A. and Antonio C. Lasaga. "Modeling the Geochemical Carbon Cycle." *Scientific American* Vol. 260 (March 1989): 74–81.

Cohn, Jeffrey P. "Gauging the Biological Impacts of the Greenhouse Effect." *BioScience* Vol. 39 (March 1989): 142–146.

Graedel, Thomas E. and Paul J. Crutzen. "The Changing Atmosphere." *Scientific American* Vol. 262 (September 1989): 58–68.

Jager, Jill. "Anticipating Climatic Change." *Environment* Vol. 6 (September 1988): 13–30.

Lovejoy, Thomas E. "Will Unexpectedly the Top Blow Off?" *BioScience* Vol. 38 (November 1988): 722–726.

Monastersky, Richard. "Global Change: the Scientific Challenge." *Science News* Vol. 135 (April 15, 1989): 232–325.

Ramanathan, V. "The Greenhouse Theory of Climate Change: A Test by an Inadvertent Global Experiment." *Science* Vol. 240 (April 15, 1988): 293–299.

Roberts, Leslie. "Is There Life After Climate Change?" *Science* Vol. 242 (November 18, 1988): 1010–1013.

Schneider, Stephen H. "The Changing Climate." *Scientific American* Vol. 261 (September 1989): 70–79.

______. "The Greenhouse Effect: Science and Policy." *Science* Vol. 243 (February 10, 1989): 771–779.

______. "Climate Modeling." *Scientific American* Vol. 256 (May 1987): 72–80.

Tangley, Laura. "Preparing for Climate Change." *BioScience* Vol. 38 (January 1988): 14–18.

GLOBAL EXTINCTION

Booth, William. "Monitoring the Fate of the Forests from Space." *Science* Vol. 243 (March 17, 1989): 1428–1429.

Clark, William C. "Managing Planet Earth." *Scientific American* Vol. 261 (September 1989): 47–54.

Colinvaux, Paul A. "The Past and Future Amazon." *Scientific American* Vol. 260 (May 1989): 102–108.

Ellis, William S. "Brazil's Imperiled Rain Forest." *National Geographic* Vol. 174 (December 1988): 772–799.

Gibbons, Boyd. "Do We Treat Our Soil Like Dirt?" *National Geographic* Vol. 166 (September 1984): 353–388.

Keyfitz, Nathan. "The Growing Human Population." *Scientific American* Vol. 261 (September 1989): 119–126.

la Reviere, J.W. Maurits. "Threats to the World's Water." *Scientific American* Vol. 262 (September 1989): 80–94.

Lewin, Roger. "Damage to Tropical Forests, or Why Were There So Many Kinds of Animals." *Science* Vol. 234 (October 10, 1986): 149–150.

______. "A Mass Extinction Without Asteroids." *Science* Vol. 234 (October 3, 1986): 14–15.

Perry, Mary J. "Assessing Marine Primary Production from Space." *BioScience* Vol. 36 (July/August 1986): 461–466.

Repetto, Robert. "Deforestation in the Tropics." *Scientific American* Vol. 262 (April 1990): 36–42.

Tangley, Laura. "Acid Rain Threatens Marine Life." *BioScience* Vol. 38 (September 1988): 538–539.

Wilson, Edward O. "Threats to Biodiversity." *Scientific American* Vol. 261 (September 1989): 108–116.

Index

E

F

G

R

S

T

U

V

W

OTHER BESTSELLERS OF RELATED INTEREST